Human Anatomy

Human Anatomy

Robin Piper

MURPHY & MOORE
www.murphy-moorepublishing.com

Human Anatomy
Robin Piper
ISBN: 978-1-63987-311-1 (Hardback)

MURPHY & MOORE

Published by Murphy & Moore Publishing,
1 Rockefeller Plaza,
New York City, NY 10020, USA

Cataloging-in-Publication Data

Human anatomy / Robin Piper.
p. cm.
Includes bibliographical references and index.
ISBN 978-1-63987-311-1
1. Human anatomy. 2. Anatomy. 3. Human biology. 4. Human body.
5. Medical sciences. I. Piper, Robin.

QM23.2 .H86 2022
611--dc23

For more information regarding Murphy & Moore Publishing and its products, please visit the publisher's website www.murphy-moorepublishing.com

Table of Contents

Preface

This book is a culmination of my many years of practice in this field. I attribute the success of this book to my support group. I would like to thank my parents who have showered me with unconditional love and support and my peers and professors for their constant guidance.

The shape and structure of human being is known as human body. It comprises of different cells, tissues and organs. The study of the form and shape of human body is known as human anatomy. It studies structures like bones, cartilage, connective tissues, organs and others that determine the shape of the body. The discipline is closely related to other fields like physiology, embryology and histology that are also used for the study of human body. Modern anatomy is concerned with the study of cellular structures of organs and tissues. It uses many techniques like computed tomography, ultrasound imaging, magnetic resonance imaging and fluoroscopy to study the body in detail. Human anatomy is an upcoming field of science that has undergone rapid development over the past few decades. Most of the topics introduced in this book cover new techniques and applications of human anatomy. It aims to serve as a resource guide for students and experts alike and contribute to the growth of the discipline.

The details of chapters are provided below for a progressive learning:

Chapter - Introduction

The structure of the human being composed of different cells that form tissues and the organ systems is termed as the human body. The study of the shape and form of the human body falls under the domain of human anatomy. This chapter has been carefully written to provide an extensive understanding of human anatomy.

Chapter - Human Body Systems

The human body system is made up of eleven major organ systems. It includes circulatory, respiratory, digestive, excretory, nervous and endocrine systems, immune, integumentary, skeletal, muscle and reproductive systems. This chapter discusses the subject of human body systems in detail.

Chapter - Human Head and Neck

The anatomy of the head and neck of human body includes the brain, bones, muscles, blood vessels, nerves, glands, nose, mouth, teeth, tongue, and throat. This chapter closely examines the human head and neck anatomy to provide an easy understanding of the subject.

Chapter - Upper Limb Anatomy

The upper limb anatomy consists of three sections namely the upper arm, forearm, and hand that extends from the shoulder joint to the fingers. It also has many nerves, arteries, veins, and muscles. The topics elaborated in this chapter will help in gaining a better perspective of upper limb anatomy.

Chapter - Lower Limb Anatomy

The upper leg, the lower leg, and the foot together comprises the lower limb anatomy. The leg consists of two long bones, the tibia and fibula, and the sesamoid bone up. The patella serves as the knee-cap and the thigh consists of a single bone, the femur. This chapter sheds light on the subject of lower limb anatomy for a thorough understanding of the subject.

Robin Piper

1
Introduction

The structure of the human being composed of different cells that form tissues and the organ systems is termed as the human body. The study of the shape and form of the human body falls under the domain of human anatomy. This chapter has been carefully written to provide an extensive understanding of human anatomy.

Human Body

Human body is the physical substance of the human organism composed of living cells and extracellular materials and organized into tissues, organs, and systems.

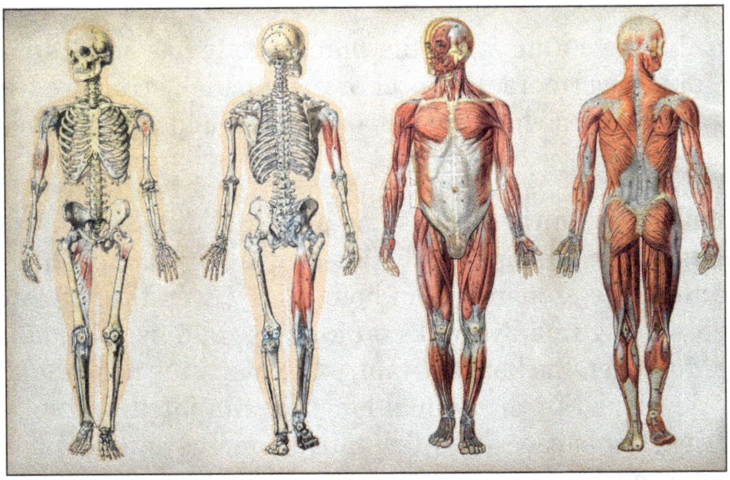

Human body; human anatomy: Vintage anatomy charts of the human body showing the skeletal and muscle systems.

Humans are, of course, animals—more particularly, members of the order Primates in the subphylum Vertebrata of the phylum Chordata. Like all chordates, the human animal has a bilaterally symmetrical body that is characterized at some point during its development by a dorsal supporting rod (the notochord), gill slits in the region of the pharynx, and a hollow dorsal nerve cord. Of these features, the first two are present only during the embryonic stage in the human; the notochord is replaced by the vertebral column, and the pharyngeal gill slits are lost completely. The dorsal nerve cord is the spinal cord in humans; it remains throughout life.

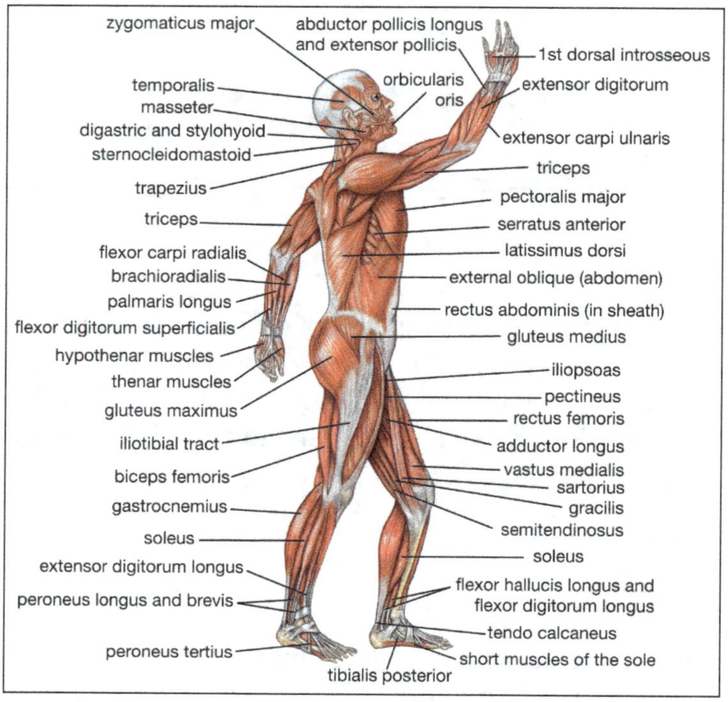

Human muscular system: lateral view. Lateral view of the human muscular system.

Characteristic of the vertebrate form, the human body has an internal skeleton that includes a backbone of vertebrae. Typical of mammalian structure, the human body shows such characteristics as hair, mammary glands, and highly developed sense organs.

Beyond these similarities, however, lie some profound differences. Among the mammals, only humans have a predominantly two-legged (bipedal) posture, a fact that has greatly modified the general mammalian body plan. (Even the kangaroo, which hops on two legs when moving rapidly, walks on four legs and uses its tail as a "third leg" when standing.) Moreover, the human brain, particularly the neocortex, is far and away the most highly developed in the animal kingdom. As intelligent as are many other mammals—such as chimpanzees and dolphins—none have achieved the intellectual status of the human species.

Chemical Composition of the Body

Chemically, the human body consists mainly of water and of organic compounds—i.e., lipids, proteins, carbohydrates, and nucleic acids. Water is found in the extracellular fluids of the body (the blood plasma, the lymph, and the interstitial fluid) and within the cells themselves. It serves as a solvent without which the chemistry of life could not take place. The human body is about 60 percent water by weight.

Lipids—chiefly fats, phospholipids, and steroids—are major structural components of the human body. Fats provide an energy reserve for the body, and fat pads also serve as insulation and shock absorbers. Phospholipids and the steroid compound cholesterol are major components of the membrane that surrounds each cell. Proteins also serve as a major structural component of the body. Like lipids, proteins are an important constituent of the cell membrane. In addition, such extracellular materials as hair and nails are composed of protein. So also is collagen, the fibrous, elastic material that makes up much of the body's skin, bones, tendons, and ligaments. Proteins also perform numerous functional roles in the body. Particularly important are cellular proteins called enzymes, which catalyze the chemical reactions necessary for life.

Carbohydrates are present in the human body largely as fuels, either as simple sugars circulating through the bloodstream or as glycogen, a storage compound found in the liver and the muscles. Small amounts of carbohydrates also occur in cell membranes, but, in contrast to plants and many invertebrate animals, humans have little structural carbohydrate in their bodies.

Nucleic acids make up the genetic materials of the body. Deoxyribonucleic acid (DNA) carries the body's hereditary master code, the instructions according to which each cell operates. It is DNA, passed from parents to offspring, that dictates the inherited characteristics of each individual human. Ribonucleic acid (RNA), of which there are several types, helps carry out the instructions encoded in the DNA.

Along with water and organic compounds, the body's constituents include various inorganic minerals. Chief among these are calcium, phosphorus, sodium, magnesium, and iron. Calcium and phosphorus, combined as calcium-phosphate crystals, form a large part of the body's bones. Calcium is also present as ions in the blood and interstitial fluid, as is sodium. Ions of phosphorus, potassium, and magnesium, on the other hand, are abundant within the intercellular fluid. All of these ions play vital roles in the body's metabolic processes. Iron is present mainly as part of hemoglobin, the oxygen-carrying pigment of the red blood cells. Other mineral constituents of the body, found in minute but necessary concentrations, include cobalt, copper, iodine, manganese, and zinc.

Organization of the Body

The cell is the basic living unit of the human body—indeed, of all organisms. The human body consists of trillions of cells, each capable of growth, metabolism, response to stimuli, and, with some exceptions, reproduction. Although there are some 200 different types of cells in the body, these can be grouped into four basic classes. These four basic cell types, together with their extracellular materials, form the fundamental tissues of the human body: (1) epithelial tissues, which cover the body's surface and line the internal organs, body cavities, and passageways; (2) muscle tissues, which are capable of contraction and form the body's musculature; (3) nerve tissues, which conduct electrical impulses and make up the nervous system; and (4) connective tissues, which are composed of widely spaced cells and large amounts of intercellular matrix and which bind together various body structures. (Bone and blood are considered specialized connective tissues, in which the intercellular matrix is, respectively, hard and liquid).

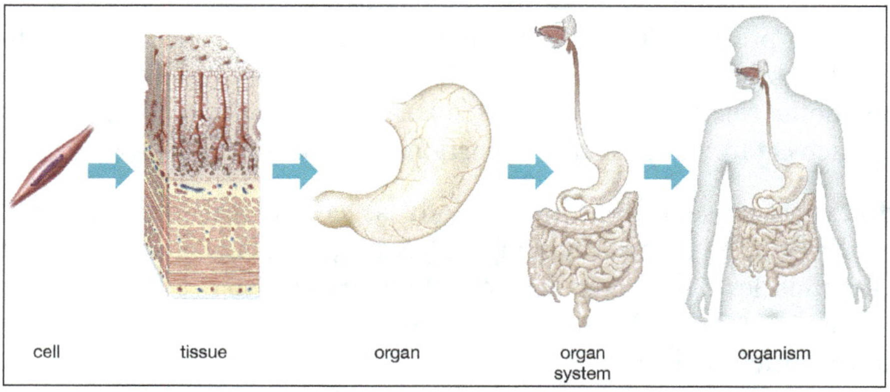

Multicelluar organism: organization.

The diagram shows five levels of organization in a multicellular organism. The most basic unit is the cell; groups of similar cells form tissues; groups of different tissues make up organs; groups of organs form organ systems; cells, tissues, organs, and organ systems combine to form a multicellular organism.

The next level of organization in the body is that of the organ. An organ is a group of tissues that constitutes a distinct structural and functional unit. Thus, the heart is an organ composed of all four tissues, whose function is to pump blood throughout the body. Of course, the heart does not function in isolation; it is part of a system composed of blood and blood vessels as well. The highest level of body organization, then, is that of the organ system.

The body includes nine major organ systems, each composed of various organs and tissues that work together as a functional unit. The chief constituents and prime functions of each system are summarized below:

- The integumentary system, composed of the skin and associated structures, protects the body from invasion by harmful microorganisms and chemicals; it also prevents water loss from the body.
- The musculoskeletal system (also referred to separately as the muscle system and the skeletal system), composed of the skeletal muscles and bones (with about 206 of the latter in adults), moves the body and protectively houses its internal organs.
- The respiratory system, composed of the breathing passages, lungs, and muscles of respiration, obtains from the air the oxygen necessary for cellular

metabolism; it also returns to the air the carbon dioxide that forms as a waste product of such metabolism.

- The circulatory system, composed of the heart, blood, and blood vessels, circulates a transport fluid throughout the body, providing the cells with a steady supply of oxygen and nutrients and carrying away waste products such as carbon dioxide and toxic nitrogen compounds.
- The digestive system, composed of the mouth, esophagus, stomach, and intestines, breaks down food into usable substances (nutrients), which are then absorbed from the blood or lymph; this system also eliminates the unusable or excess portion of the food as fecal matter.
- The excretory system, composed of the kidneys, ureters, urinary bladder, and urethra, removes toxic nitrogen compounds and other wastes from the blood.
- The nervous system, composed of the sensory organs, brain, spinal cord, and nerves, transmits, integrates, and analyzes sensory information and carries impulses to effect the appropriate muscular or glandular responses.
- The endocrine system, composed of the hormone-secreting glands and tissues, provides a chemical communications network for coordinating various body processes.
- The reproductive system, composed of the male or female sex organs, enables reproduction and thereby ensures the continuation of the species.

Basic Form and Development

In general structure, the human body follows a plan that can be described as a cylinder enclosing two tubes and a rod. This body plan is most clearly evident in the embryo; by birth, the plan is apparent only in the trunk region—i.e., in the thorax and abdomen.

The body wall forms the cylinder. The two tubes are the ventrally located alimentary canal (i.e., the digestive tract) and the dorsally located neural tube (i.e., the spinal cord). Between the tubes lies the rod—the notochord in the embryo, which becomes the vertebral column prior to birth. (The terms dorsal and ventral refer respectively to the back and the front, or belly, of an animal).

Within the embryo, the essential body parts are:

- The outer enclosing epidermal membrane (in the embryo called ectoderm).
- The dorsal neural tube.
- The supporting notochord.

- The ventral alimentary tube, which becomes the lining of the stomach and intestine (in the embryo called endoderm).
- The intermediate mass (in the embryo called mesoderm).
- A rather fluid tissue that fills the interspaces derived from the mesoderm and in the embryo called mesenchyme. Everything in the body derives from one of these six embryonic parts.

The mesoderm constitutes a considerable pad of tissue on each side of the embryo, extending all the way from the back to the front sides of the body wall. It is hollow, for a cleftlike space appears in it on each side. These are the right and left body cavities. In the dorsal part of the body they are temporary; in the ventral part they become permanent, forming the two pleural cavities, which house the lungs; the peritoneal cavity, which contains the abdominal organs; and the pericardial cavity, which encloses the heart. The dorsal part of the mesoderm becomes separated from the ventral mesoderm and divides itself into serial parts like a row of blocks, 31 on each side. These mesodermal segments grow in all directions toward the epidermal membrane. They form bones, muscles, and the deeper, leathery part of the skin. Dorsally they form bony arches protecting the spinal cord, and ventrally the ribs protecting the alimentary canal and heart. Thus they form the body wall and the limbs—much the weightier part of the body. They give the segmental character to the body wall in neck and trunk, and, following their lead, the spinal cord becomes correspondingly segmented. The ventral mesoderm is not so extensive; it remains near the alimentary tube and becomes the continuous muscle layer of the stomach and intestine. It also forms the lining of the body cavities, the smooth, shining, slippery pleura and peritoneum. The mesenchyme forms blood and lymph vessels, the heart, and the loose cells of connective tissues.

The neural tube itself is formed from the ectoderm at a very early stage. Anteriorly (i.e., toward the head) it extends above the open end of the cylinder and is enlarged to form the brain. It is not in immediate contact with the epidermis, for the dorsal mesoderm grows up around it and around the roots of the cranial nerves as a covering, separating the brain from the epidermis. Posteriorly the neural tube terminates in the adult opposite the first lumbar vertebra.

If the cylindrical body wall is followed headward, it is found to terminate ventrally as the tongue, dorsally in the skull around the brain, ears, and eyes. There is a considerable interval between eyes and tongue. This is occupied partly by a deep depression of the epidermis between them, which dips in to join the alimentary tube (lining of the mouth). Posteriorly the ventral body wall joins the dorsal at the tailbone (coccyx), thus terminating the body cavities.

Headward, the alimentary tube extends up in front of the notochord and projects above the upper part of the body wall (tongue) and in front of and below the brain to join the epidermal depression. From the epidermal depression are formed the teeth and most of the mouth lining; from the upper end of the alimentary canal are formed the pharynx, larynx, trachea, and lungs. The alimentary canal at its tail end splits longitudinally into two tubes—an anterior and a posterior. The anterior tube becomes the bladder, urethra, and, in the female, the lining of the vagina, where it joins a depression of the ectoderm. The posterior (dorsal) tube becomes the rectum and ends just in front of the coccyx by joining another ectodermal depression (the anus).

Human Anatomy

Human anatomy can be precisely defined as a complementary basic medical science, which deals with the scientific study of morphology of human body. In simpler words, human anatomy is the study of structure of human body.

Subdivisions of Human Anatomy

There are two main levels of structure of human body (as well as every other thing): macroscopic level and microscopic level. For each of the two levels there is a separate subdivision of anatomy. The one dealing with macroscopic level is known as gross anatomy and the other which deals with microscopic level is called microscopic anatomy or histology.

Gross Anatomy

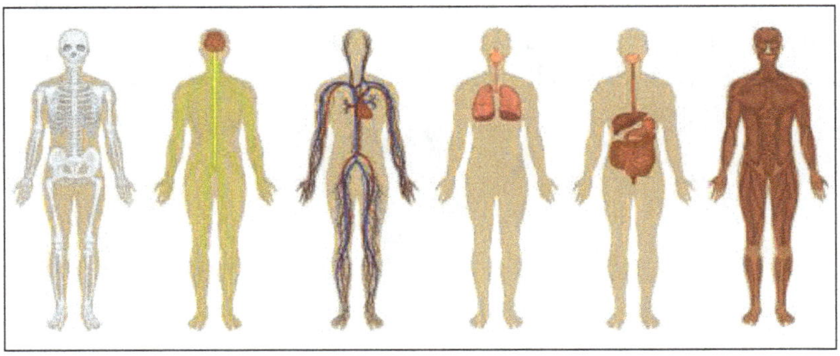

In gross anatomy, structure of human body is studied as seen by naked eye. There are two approaches for gross anatomy: Regional approach and Systemic approach.

- In regional approach, human body is divided into regions like arm, forearm, wrist etc. and all structures within that region are studied simultaneously.
- Whereas in systemic approach, the body is divided into organ systems such as skeletal system, muscular system, nervous system etc. and structures belonging to one system are studied together.

Histology

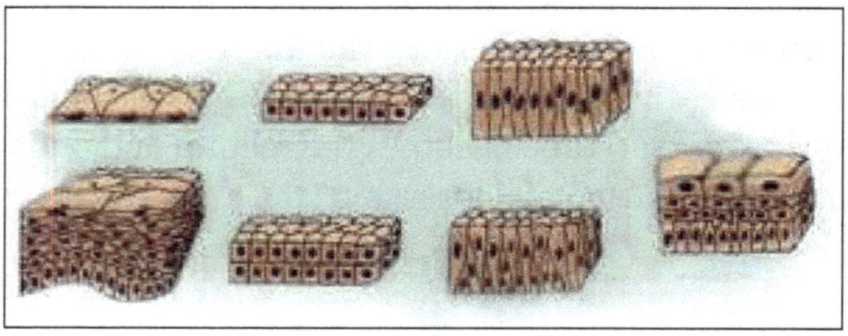

Histology or microscopic anatomy is the study of the structure of various organs and tissues of human body under a microscope. The understanding of the ultra-structure helps understand the tissues and organs in a better way.

Basic Anatomy

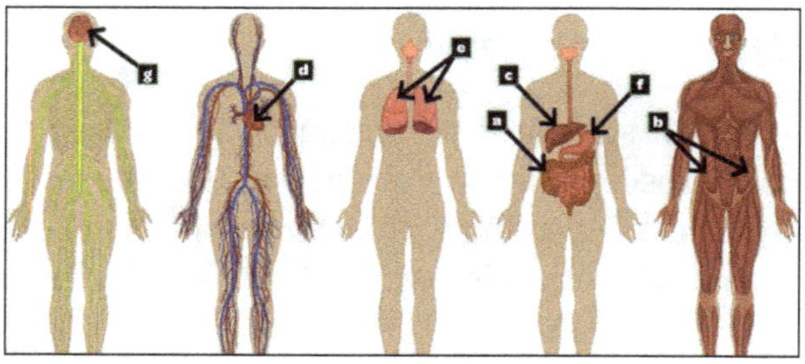

In addition to the main subdivisions of human anatomy, a third branch, called basic anatomy, is considered of significant importance. It explains the basic terms and definitions used in the study of gross as well as microscopic anatomy. Thus it provides an introduction to anatomy and tells how to study it.

2

Human Body Systems

The human body system is made up of eleven major organ systems. It includes circulatory, respiratory, digestive, excretory, nervous and endocrine systems, immune, integumentary, skeletal, muscle and reproductive systems. This chapter discusses the subject of human body systems in detail.

Skeletal System

The skeletal system provides support and protection for the body's internal organs and gives the muscles a point of attachment. Humans have an endoskeleton, where our bones lie underneath our skin and muscles. In other animals, such as insects, there is an exoskeleton on the outside of the body.

In humans, the skeletal system consists of bones, joints and associated cartilages. An adult human has 206 bones in their body and variety of different joints.

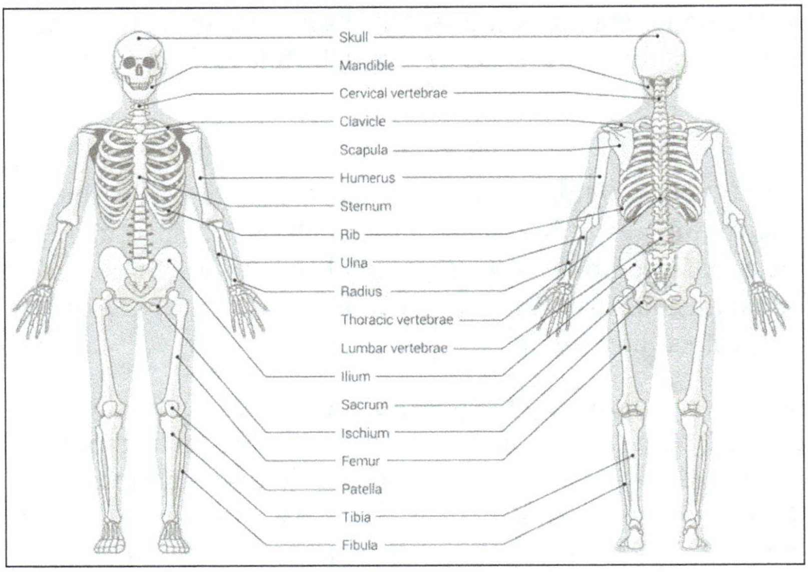

Skeletal System Diagram.

The human skeleton can be divided into two components: the axial skeleton and the appendicular skeleton. The axial skeleton is formed around the central axis of the body and thus includes the skull, spine, and ribcage. It protects the brain, spinal cord, heart, lungs, esophagus and major sense organs like the eyes, ears, nose, and tongue. The appendicular skeleton is related to the limbs and consists of the bones of the arms and legs, as well as the shoulder and hip girdles.

Skeletal System Function

Support

The first and most apparent function of the skeletal system is to provide a framework for the body. The presence of a firm bony skeleton allows the organism to have a distinctive shape adapted towards a particular lifestyle. For instance, in a fast-moving animal like the cheetah, the skeleton contains long, thin limb bones and an extremely flexible spine. The structure of the skeleton also allows it to absorb the impact of running at high speeds.

The bones of birds are hollow, light and create a streamlined body adapted for flight. Many animals even have sexual dimorphism in their skeletons. In humans, while this dimorphism is fairly limited, there are differences in the angle of the pelvic bones, to accommodate pregnancy.

Integration with the Muscular System

The skeletal system also provides an important form of attachments to the muscular system. Bones and exoskeletons are hard and do not bend or move when muscles are flexed. This means that the contraction of muscle cells will lead to the shortening of muscles, while the bone retains its shape. This basic structure allows muscles to move different parts of the body, using forces generated while pulling on the skeletal system.

Protection

The next obvious function of the skeletal system is the role it plays protecting the fragile internal organs. In humans, this is seen in the skull, which surrounds the brain completely. It is also exhibited by the ribcage, which surrounds the lungs and heart but still allows for expansion. Even invertebrates like snails and prawns often have hard exoskeletons to protect themselves from predators.

The rigid endoskeleton allows the body to rise up above the ground or stand upright, and bears the weight of the organism, and provides the scaffolding for movement. Muscles generate the force required to move bones at joints. Muscle fibers contain actin and myosin, two protein filaments that can slide past each other to change the length of the muscle. When a nerve impulse arrives at the neuromuscular junction, it signals the muscle to contract. The force generated by the contracting muscle either pulls two bones together or apart, based on the nature of the interaction between the muscle and joint.

Blood Cell Production

The central part of a bone contains the bone marrow, the primary site for blood cell production in adult humans. There are two types of bone marrow in adults. Around 50% is red bone marrow containing hematopoietic stem cells and supportive tissue. The rest is yellow bone marrow made of fat and its proportion increases with age. Bone marrow will revert to a higher proportion of red marrow if the body suffers an injury and needs to create more red blood cells. The bone marrow composition also changes during pregnancy and lactation in mammals. Over the course of gestation, blood volume increases by about 1.5 liters, and even the concentration of red blood cells and white blood cells increase.

Production of other Cell Types

In addition to producing red blood cells, bone marrow within the skeletal system is the production site of a number of other cells. These include lymphocytes, which are immune cells that travel the lymphatic system. In addition to providing immune functions, the skeletal system is also responsible for hosting stem cells which can differentiate into muscle cells, cartilage-producing cells, and cells that create bone (osteoblasts).

Osteoblasts in bone also have an endocrine function, secreting a hormone called osteocalcin. It requires vitamin K to be synthesized and is an anabolic hormone. It mediates an increase in insulin levels and increases the sensitivity of the body to insulin. Osteocalcin contributes to an increase in bone mass and bone mineralization.

Storing Minerals

The bones of the skeletal system act as a storehouse for calcium ions, changing the quantum of mineralized deposits within bones to maintain plasma calcium ion concentration within a narrow range. Calcium ions can affect crucial sodium ion channels in the plasma membrane of every cell, thereby affecting overall homeostasis.

For this reason, changes to the concentration of calcium ions have particularly adverse effects on excitable cells in the nervous system, and in cardiac, skeletal and smooth muscle. Different interacting hormones maintain the balance of calcium ions in the plasma and bones, especially the parathyroid hormone secreted from the parathyroid glands in the neck.

Skeletal System Parts

The anatomy of the skeletal system is complex, and it includes hundreds of bones in the human body. The anatomy of the system varies widely between organisms, as evolution has selected for various adaptations in certain species which change the structure and function of their bones.

- Bone: Bones serve a variety of functions, but the most important is supporting movement of the limbs and body. Two bones or cartilages are held together at a joint through tough connective tissues called ligaments. Muscles are securely attached to bones through flexible but inelastic connective tissue called tendons. Muscles, joints, tendons, and ligaments are part of the intricate machinery that allows the movement of different bones.
- Joints: Functionally, joints can be divided into three classes based on the range of movement they allow in the associated bones. Immovable joints are formed when two bones are held together by fibrous connective tissue with no synovial fluid. These kinds of joints hold the bones of the cranium together. Partially movable joints are also called cartilaginous joints and are present in the spine and ribs. The third type of joints are called synovial joints and have a fluid-filled synovial cavity that allows the interfacing bones the largest range of movement. Based on the structure of the synovial joints, they can be classified into 6 types, including the hinge joints of the fingers and the ball and socket joints of the hips and shoulders.

Cellular Composition

Each bone is made of complex sets of cells, tissues and a specialized extracellular matrix. The two main cell types are called osteoblasts and osteoclasts with mostly opposing functions. While osteoblasts are involved in the formation of bone, osteoclasts are associated with a reduction in bone mass. The extracellular matrix of the bone consists of collagen and other organic fibers as well as the inorganic component containing calcium salts such as hydroxyapatite. In the interior of bones, a soft tissue called the bone marrow plays an important role in immunity and hematopoiesis. The bone is also richly supplied with nerves and blood vessels.

Muscular System

The muscular system is an organ system consisting of skeletal, smooth and cardiac muscles. It permits movement of the body, maintains posture and circulates blood throughout the body. The muscular systems in vertebrates are controlled through the nervous system although some muscles (such as the cardiac muscle) can be completely autonomous. Together with the skeletal system, it forms the musculoskeletal system, which is responsible for movement of the human body.

Muscles

There are three distinct types of muscles: skeletal muscles, cardiac or heart muscles, and smooth (non-striated) muscles. Muscles provide strength, balance, posture, movement and heat for the body to keep warm.

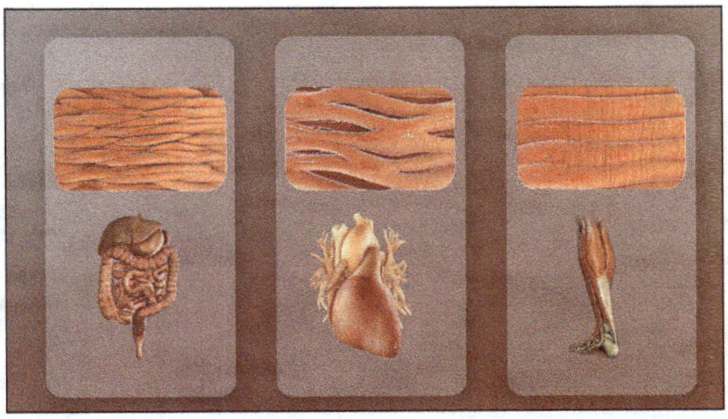

Three distinct types of muscles (L to R): Smooth (non-striated) muscles in internal organs, cardiac or heart muscles and skeletal muscles.

Skeletal Muscle

Skeletal muscles, like other striated muscles, are composed of myocytes, or muscle fibers, which are in turn composed of myofibrils, which are composed of sarcomeres, the basic building block of striated muscle tissue. Upon stimulation by an action potential, skeletal muscles perform a coordinated contraction by shortening each sarcomere. The best proposed model for understanding contraction is the sliding filament model of muscle contraction. Within the sarcomere, actin and myosin fibers overlap in a contractile motion towards each other. Myosin filaments have club-shaped heads that project toward the actin filaments.

Larger structures along the myosin filament called myosin heads are used to provide attachment points on binding sites for the actin filaments. The myosin heads move in a coordinated style; they swivel toward the center of the sarcomere, detach and then reattach to the nearest active site of the actin filament. This is called a ratchet type drive system.

This process consumes large amounts of adenosine triphosphate (ATP), the energy source of the cell. ATP binds to the cross bridges between myosin heads and actin filaments. The release of energy powers the swiveling of the myosin head. When ATP is used, it becomes adenosine diphosphate (ADP), and since muscles store little ATP, they must continuously replace the discharged ADP with ATP. Muscle tissue also contains a stored supply of a fast acting recharge chemical, creatine phosphate, which when necessary can assist with the rapid regeneration of ADP into ATP.

Calcium ions are required for each cycle of the sarcomere. Calcium is released from the sarcoplasmic reticulum into the sarcomere when a muscle is stimulated to contract. This calcium uncovers the actin binding sites. When the muscle no longer needs to contract, the calcium ions are pumped from the sarcomere and back into storage in the sarcoplasmic reticulum.

There are approximately 639 skeletal muscles in the human body.

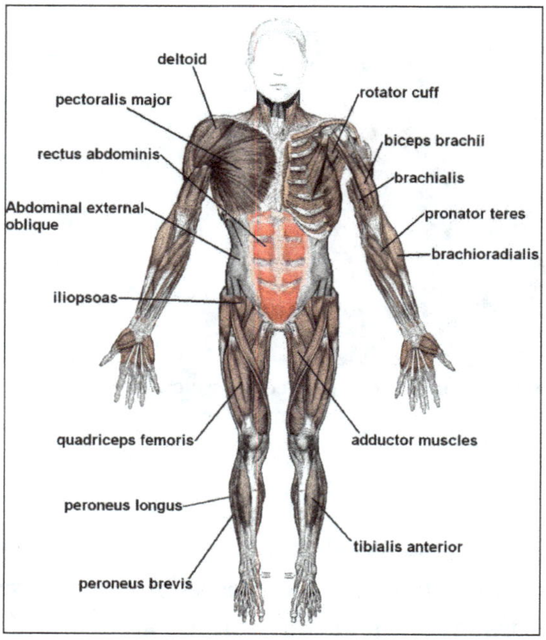

Skeletal muscles, viewed from the front.

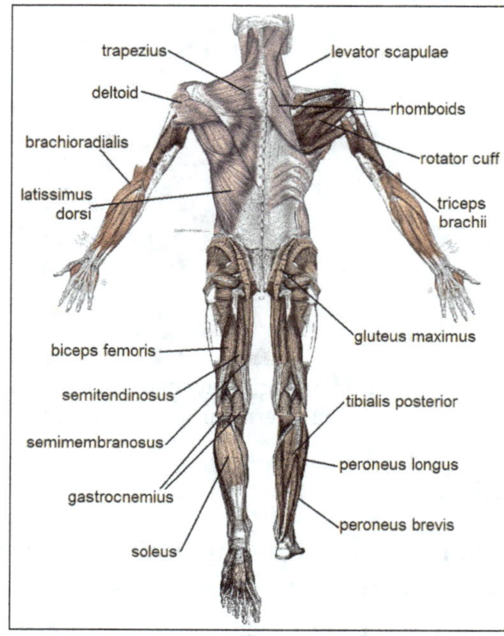

Skeletal muscles, viewed from the back.

Cardiac Muscle

Heart muscles are distinct from skeletal muscles because the muscle fibers are laterally connected to each other. Furthermore, just as with smooth muscles, their movement is involuntary. Heart muscles are controlled by the sinus node influenced by the autonomic nervous system.

Smooth Muscle

Smooth muscles are controlled directly by the autonomic nervous system and are involuntary, meaning that they are incapable of being moved by conscious thought. Functions such as heartbeat and lungs (which are capable of being willingly controlled, be it to a limited extent) are involuntary muscles but are not smooth muscles.

Physiology

Contraction

Neuromuscular junctions are the focal point where a motor neuron attaches to a muscle. Acetylcholine, (a neurotransmitter used in skeletal muscle contraction) is released from the axon terminal of the nerve cell when an action potential reaches the microscopic junction called a synapse. A group of chemical messengers cross the synapse and stimulate the formation of electrical changes, which are produced in the muscle cell when the acetylcholine binds to receptors on its surface. Calcium is released from its storage area in the cell's sarcoplasmic reticulum. An impulse from a nerve cell causes calcium release and brings about a single, short muscle contraction called a muscle twitch. If there is a problem at the neuromuscular junction, a very prolonged contraction may occur, such as the muscle contractions that result from tetanus. Also, a loss of function at the junction can produce paralysis.

Skeletal muscles are organized into hundreds of motor units, each of which involves a motor neuron, attached by a series of thin finger-like structures called axon terminals. These attach to and control discrete bundles of muscle fibers. A coordinated and fine tuned response to a specific circumstance will involve controlling the precise number of motor units used. While individual muscle units contract as a unit, the entire muscle can contract on a predetermined basis due to the structure of the motor unit. Motor unit coordination, balance, and control frequently come under the direction of the cerebellum of the brain. This allows for complex muscular coordination with little conscious effort, such as when one drives a car without thinking about the process.

Aerobic and Anaerobic Muscle Activity

At rest, the body produces the majority of its ATP aerobically in the mitochondria without producing lactic acid or other fatiguing byproducts. During exercise, the method of ATP production varies depending on the fitness of the individual as well as the duration and intensity of exercise. At lower activity levels, when exercise continues for a long duration (several minutes or longer), energy is produced aerobically by combining oxygen with carbohydrates and fats stored in the body.

During activity that is higher in intensity, with possible duration decreasing as intensity increases, ATP production can switch to anaerobic pathways, such as the use of the creatine phosphate and the phosphagen system or anaerobic glycolysis. Aerobic ATP production is biochemically much slower and can only be used for long-duration, low-intensity exercise, but produces no fatiguing waste products that cannot be removed immediately from the sarcomere and the body, and it results in a much greater number of ATP molecules per fat or carbohydrate molecule. Aerobic training allows the oxygen delivery system to be more efficient, allowing aerobic metabolism to begin quicker. Anaerobic ATP production produces ATP much faster and allows near-maximal intensity exercise, but also produces significant amounts of lactic acid which renders high-intensity exercise unsustainable for more than several minutes. The phosphagen system is also anaerobic. It allows for the highest levels of exercise intensity, but intramuscular stores of phosphocreatine are very limited and can only provide energy for exercises lasting up to ten seconds. Recovery is very quick, with full creatine stores regenerated within five minutes.

Nervous System

The nervous system is a highly complex part of an animal that coordinates its actions and sensory information by transmitting signals to and from different parts of its body. The nervous system detects environmental changes that impact the body, then works in tandem with the endocrine system to respond to such events. Nervous tissue first arose in wormlike organisms about 550 to 600 million years ago. In vertebrates it consists of two main parts, the central nervous system (CNS) and the peripheral nervous system (PNS). The CNS consists of the brain and spinal cord. The PNS consists mainly of nerves, which are enclosed bundles of the long fibers or axons, that connect the CNS to every other part of the body. Nerves that transmit signals from the brain are called motor or efferent nerves, while those nerves that transmit information from the body to the CNS are called sensory or afferent. Spinal nerves serve both functions and are called mixed nerves. The PNS is divided into three separate subsystems, the somatic, autonomic, and enteric nervous systems. Somatic nerves mediate voluntary movement. The autonomic nervous system is further subdivided into the sympathetic and the parasympathetic nervous systems. The sympathetic nervous system is activated in cases of emergencies to mobilize energy, while the parasympathetic nervous system is activated when organisms are in a relaxed state. The enteric nervous system functions to control the gastrointestinal system. Both autonomic and enteric nervous systems function involuntarily. Nerves that exit from the cranium are called cranial nerves while those exiting from the spinal cord are called spinal nerves.

At the cellular level, the nervous system is defined by the presence of a special type of cell, called the neuron, also known as a "nerve cell". Neurons have special structures that allow them to send signals rapidly and precisely to other cells. They send these signals in the form of electrochemical waves traveling along thin fibers called axons, which cause chemicals called neurotransmitters to be released at junctions called synapses. A cell that receives a synaptic signal from a neuron may be excited, inhibited, or otherwise modulated. The connections between neurons can form neural pathways, neural circuits, and larger networks that generate an organism's perception of the world and determine its behavior. Along with neurons, the nervous system contains other specialized cells called glial cells (or simply glia), which provide structural and metabolic support.

Nervous systems are found in most multicellular animals, but vary greatly in complexity. The only multicellular animals that have no nervous system at all are sponges, placozoans, and mesozoans, which have very simple body plans. The nervous systems of the radially symmetric organisms ctenophores (comb jellies) and cnidarians (which include anemones, hydras, corals and jellyfish) consist of a diffuse nerve net. All other animal species, with the exception of a few types of worm, have a nervous system containing a brain, a central cord (or two cords running in parallel), and nerves radiating from the brain and central cord. The size of the nervous system ranges from a few hundred cells in the simplest worms, to around 300 billion cells in African elephants.

The central nervous system functions to send signals from one cell to others, or from one part of the body to others and to receive feedback. Malfunction of the nervous system can occur as a result of genetic defects, physical damage due to trauma or toxicity, infection or simply of ageing. The medical specialty of neurology studies disorders of the nervous system and looks for interventions that can prevent or treat them. In the peripheral nervous system, the most common problem is the failure of nerve conduction, which can be due to different causes including diabetic neuropathy and demyelinating disorders such as multiple sclerosis and amyotrophic lateral sclerosis. Neuroscience is the field of science that focuses on the study of the nervous system.

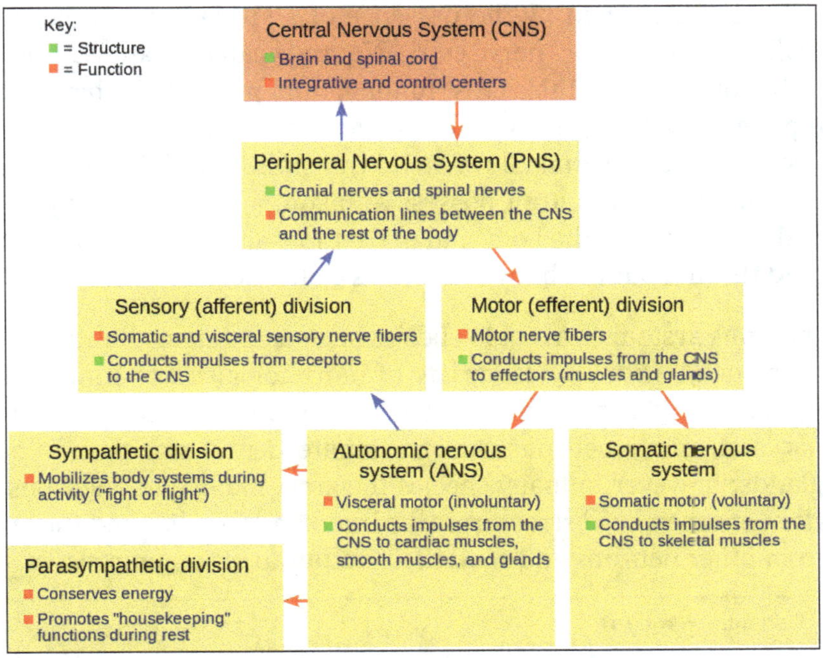

Diagram showing the major divisions of the vertebrate nervous system.

Structure

The nervous system derives its name from nerves, which are cylindrical bundles of fibers (the axons of neurons), that emanate from the brain and spinal cord, and branch repeatedly to innervate every part of the body.

A microscopic examination shows that nerves consist primarily of axons, along with different membranes that wrap around them and segregate them into fascicles. The neurons that give rise to nerves do not lie entirely within the nerves themselves—their cell bodies reside within the brain, spinal cord, or peripheral ganglia.

All animals more advanced than sponges have nervous systems. However, even sponges, unicellular animals, and non-animals such as slime molds have cell-to-cell signalling mechanisms that are precursors to those of neurons. In radially symmetric animals such as the jellyfish and hydra, the nervous system consists of a nerve net, a diffuse network of isolated cells. In bilaterian animals, which make up the great majority of existing species, the nervous system has a common structure that originated early in the Ediacaran period, over 550 million years ago.

Cells

The nervous system contains two main categories or types of cells: neurons and glial cells.

Neurons

The nervous system is defined by the presence of a special type of cell—the neuron (sometimes called "neurone" or "nerve cell"). Neurons can be distinguished from other cells in a number of ways, but their most fundamental property is that they communicate with other cells via synapses, which are membrane-to-membrane junctions containing molecular machinery that allows rapid transmission of signals, either electrical or chemical. Many types of neuron possess an axon, a protoplasmic protrusion that can extend to distant parts of the body and make thousands of synaptic contacts; axons typically extend throughout the body in bundles called nerves.

Even in the nervous system of a single species such as humans, hundreds of different types of neurons exist, with a wide variety of morphologies and functions. These include sensory neurons that transmute physical stimuli such as light and sound into neural signals, and motor neurons that transmute neural signals into activation of muscles or glands; however in many species the great majority of neurons participate in the formation of centralized structures (the brain and ganglia) and they receive all of their input from other neurons and send their output to other neurons.

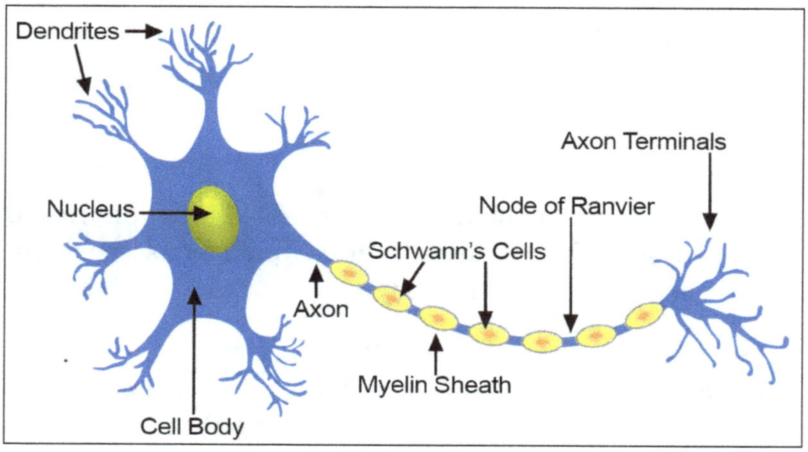

Structure of a Typical Neuron.

Glial Cells

Glial cells are non-neuronal cells that provide support and nutrition, maintain homeostasis, form myelin, and participate in signal transmission in the nervous system.

In the human brain, it is estimated that the total number of glia roughly equals the number of neurons, although the proportions vary in different brain areas. Among the most important functions of glial cells are to support neurons and hold them in place; to supply nutrients to neurons; to insulate neurons electrically; to destroy pathogens and remove dead neurons; and to provide guidance cues directing the axons of neurons to their targets. A very important type of glial cell (oligodendrocytes in the central nervous system, and Schwann cells in the peripheral nervous system) generates layers of a fatty substance called myelin that wraps around axons and provides electrical insulation which allows them to transmit action potentials much more rapidly and efficiently. Recent findings indicate that glial cells, such as microglia and astrocytes, serve as important resident immune cells within the central nervous system.

Respiratory System

Respiration is the act of breathing:

- Inhaling: The act of breathing in oxygen.
- Exhaling: The act of breathing out carbon dioxide.

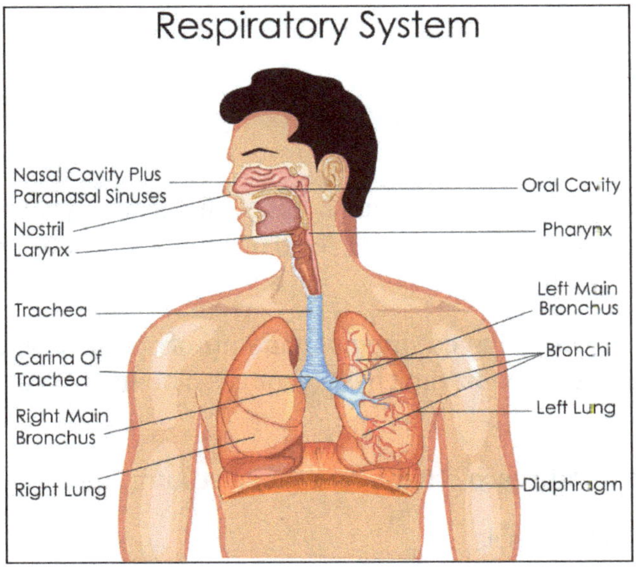

The respiratory system is made up of the organs included in the exchange of oxygen and carbon dioxide. These are the parts:

- Nose.
- Mouth.

- Throat (pharynx).
- Voice box (larynx).
- Windpipe (trachea).
- Large airways (bronchi).
- Lungs.

The upper respiratory tract is made up of the:

- Nose.
- Nasal cavity.
- Sinuses.
- Larynx.
- Trachea.

The lower respiratory tract is made up of the:

- Lungs.
- Bronchi and bronchioles.
- Air sacs (alveoli).

Lungs

The lungs take in oxygen. The cells of your body's cells need oxygen to live and carry out their normal functions. The lungs also get rid of carbon dioxide, a waste product of the cells.

The lungs are a pair of cone-shaped organs made up of spongy, pinkish-gray tissue. They take up most of the space in the chest (thorax). The lungs are surrounded by a membrane (pleura).

The lungs are separated from each other by the mediastinum, an area that contains the:

- Heart and its large vessels.
- Trachea.
- Esophagus.
- Thymus gland.
- Lymph nodes.

The right lung has three sections, called lobes. The left lung has two lobes. When you breathe:

- Air enters your body through your nose or mouth.
- Air then travels down the throat through the larynx and trachea.
- Air goes into the lungs through tubes called main-stem bronchi.

One main-stem bronchus leads to the right lung and one to the left lung:

- In the lungs, the main-stem bronchi divide into smaller bronchi.
- The smaller bronchi divide into even smaller tubes (bronchioles).
- Bronchioles end in tiny air sacs (alveoli) where the exchange of oxygen and carbon dioxide occurs.

Cardiovascular System

The cardiovascular system is the transport system of the body. This system has three main components: the heart, the blood vessel and the blood itself. The heart is the system's pump and the blood vessels are like the delivery routes. Blood can be thought of as a fluid which contains the oxygen and nutrients the body needs and carries the wastes which need to be removed. The following information describes the structure and function of the heart and the cardiovascular system as a whole.

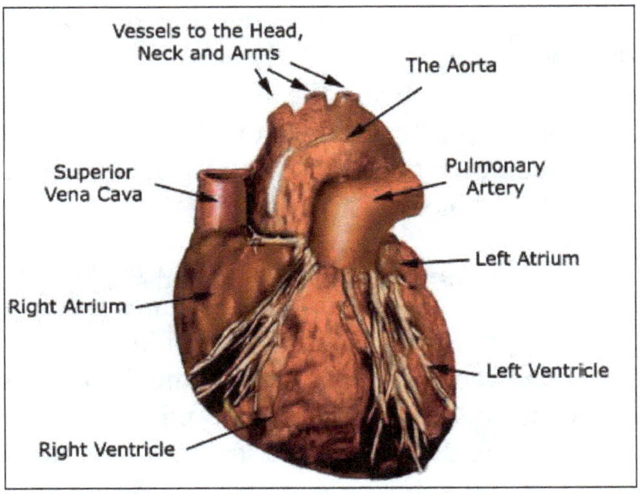

Structure and Function of the Heart

- Function and location of the heart: The heart's job is to pump blood around the body. The heart is located in between the two lungs. It lies left of the middle of the chest.

- Structure of the heart: The heart is a muscle about the size of a fist, and is roughly cone-shaped. It is about 12cm long, 9cm across the broadest point and about 6cm thick. The pericardium is a fibrous covering which wraps around the whole heart. It holds the heart in place but allows it to move as it beats. The wall of the heart itself is made up of a special type of muscle called cardiac muscle.
- Chambers of the heart: The heart has two sides, the right side and the left side. The heart has four chambers. The left and right side each have two chambers, a top chamber and a bottom chamber. The two top chambers are known as the left and right atria (singular: atrium). The atria receive blood from different sources. The left atrium receives blood from the lungs and the right atrium receives blood from the rest of the body. The bottom two chambers are known as the left and right ventricles. The ventricles pump blood out to different parts of the body. The right ventricle pumps blood to the lungs while the left ventricle pumps out blood to the rest of the body. The ventricles have much thicker walls than the atria which allows them to perform more work by pumping out blood to the whole body.
- Blood vessels: Blood Vessel are tubes which carry blood. Veins are blood vessels which carry blood from the body back to the heart. Arteries are blood vessels which carry blood from the heart to the body. There are also microscopic blood vessels which connect arteries and veins together called capillaries. There are a few main blood vessels which connect to different chambers of the heart. The aorta is the largest artery in our body. The left ventricle pumps blood into the aorta which then carries it to the rest of the body through smaller arteries. The pulmonary trunk is the large artery which the right ventricle pumps into. It splits into pulmonary arteries which take the blood to the lungs. The pulmonary veins take blood from the lungs to the left atrium. All the other veins in our body drain into the inferior vena cava (IVC) or the superior vena cava (SVC). These two large veins then take the blood from the rest of the body into the right atrium.
- Valves: Valves are fibrous flaps of tissue found between the heart chambers and in the blood vessels. They are rather like gates which prevent blood from flowing in the wrong direction. They are found in a number of places. Valves between the atria and ventricles are known as the right and left atrioventricular valves, otherwise known as the tricuspid and mitral valves respectively. Valves between the ventricles and the great arteries are known as the semilunar valves. The aortic valve is found at the base of the aorta, while the pulmonary valve is found the base of the pulmonary trunk. There are also many valves found in veins throughout the body. However, there are no valves found in any of the other arteries besides the aorta and pulmonary trunk.

The cardiovascular system refers to the heart, blood vessels and the blood. Blood contains oxygen and other nutrients which your body needs to survive. The body takes these essential nutrients from the blood. At the same time, the body dumps waste products like carbon dioxide, back into the blood, so they can be removed. The main function of the cardiovascular system is therefore to maintain blood flow to all parts of the body, to allow it to survive. Veins deliver used blood from the body back to the heart. Blood in the veins is low in oxygen (as it has been taken out by the body) and high in carbon dioxide (as the body has unloaded it back into the blood). All the veins drain into the superior and inferior vena cava which then drains into the right atrium. The right atrium pumps blood into the right ventricle. Then the right ventricle pumps blood to the pulmonary trunk, through the pulmonary arteries and into the lungs. In the lungs the blood picks up oxygen that we breathe in and gets rid of carbon dioxide, which we breathe out. The blood is becomes rich in oxygen which the body can use. From the lungs, blood drains into the left atrium and is then pumped into the left ventricle. The left ventricle then pumps this oxygen-rich blood out into the aorta which then distributes it to the rest of the body through other arteries. The main arteries which branch off the aorta and take blood to specific parts of the body are:

- Carotid arteries, which take blood to the neck and head.
- Coronary arteries, which provide blood supply to the heart itself.
- Hepatic artery, which takes blood to the liver with branches going to the stomach.
- Mesenteric artery, which takes blood to the intestines.
- Renal arteries, which takes blood to the kidneys.
- Femoral arteries, which take blood to the legs.

The body is then able to use the oxygen in the blood to carry out its normal functions. This blood will again return back to the heart through the veins and the cycle continues.

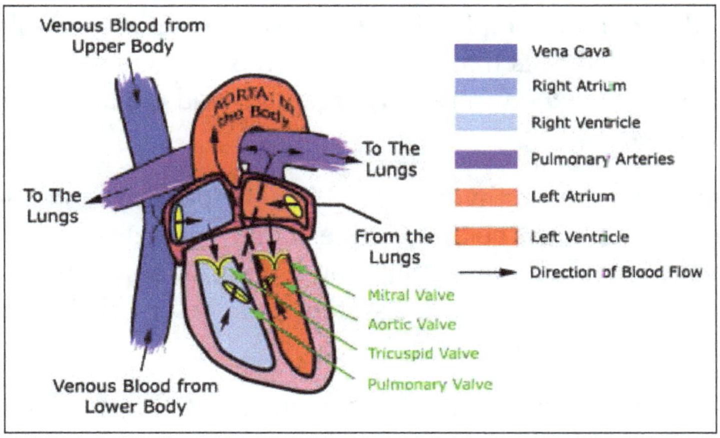

Lymphatic System

The lymphatic system consists of lymph vessels, ducts, nodes, and other tissues. Around 2 liters of fluid leak from the cardiovascular system into body tissues every day. The lymphatic system is a network of vessels that collect these fluids, or lymph. Lymph is a clear fluid that is derived from blood plasma.

The lymph vessels form a network of branches that reach most of the body's tissues. They work in a similar way to the blood vessels. The lymph vessels work with the veins to return fluid from the tissues. Unlike blood, the lymphatic fluid is not pumped but squeezed through the vessels when we use our muscles. The properties of the lymph vessel walls and the valves help control the movement of lymph. However, like veins, lymphatic vessels have valves inside them to stop fluid from flowing back in the wrong direction.

Lymph is drained progressively towards larger vessels until it reaches the two main channels, the lymphatic ducts in our trunk. From there, the filtered lymph fluid returns to the blood in the veins. The vessels branch through junctions called lymph nodes. These are often referred to as glands, but they are not true glands as they do not form part of the endocrine system.

In the lymph nodes, immune cells assess for foreign material, such as bacteria, viruses, or fungus. Lymph nodes are not the only lymphatic tissues in the body. The tonsils, spleen, and thymus gland are also lymphatic tissues.

- Tonsils: In the back of the mouth, there are tonsils. These produce lymphocytes, a type of white blood cell, and antibodies. They have a strategic position, hanging down from a ring forming the junction between the mouth and pharynx. This enables them to protect against inhaled and swallowed foreign bodies. The tonsils are the tissues affected by tonsillitis.
- Spleen: The spleen is not connected to the lymphatic system in the same way as lymph nodes, but it is lymphoid tissue. This means it plays a role in the production of white blood cells that form part of the immune system. Its other major role is to filter the blood to remove microbes and old and damaged red blood cells and platelets.
- The thymus gland: The thymus gland is a lymphatic organ and an endocrine gland that is found just behind the sternum. It secretes hormones and is crucial in the production, maturation, and differentiation of immune T cells. It is active in developing the immune system from before birth and through childhood.
- The bone marrow: Bone marrow is not lymphatic tissue, but it can be considered

part of the lymphatic system because it is here that the B cell lymphocytes of the immune system mature.

- Liver of a fetus: During gestation, the liver of a fetus is regarded as part of the lymphatic system as it plays a role in lymphocyte development.

Development

Lymphatic tissues begin to develop by the end of the fifth week of embryonic development. Lymphatic vessels develop from lymph sacs that arise from developing veins, which are derived from mesoderm.

The first lymph sacs to appear are the paired jugular lymph sacs at the junction of the internal jugular and subclavian veins. From the jugular lymph sacs, lymphatic capillary plexuses spread to the thorax, upper limbs, neck and head. Some of the plexuses enlarge and form lymphatic vessels in their respective regions. Each jugular lymph sac retains at least one connection with its jugular vein, the left one developing into the superior portion of the thoracic duct.

The next lymph sac to appear is the unpaired retroperitoneal lymph sac at the root of the mesentery of the intestine. It develops from the primitive vena cava and mesonephric veins. Capillary plexuses and lymphatic vessels spread from the retroperitoneal lymph sac to the abdominal viscera and diaphragm. The sac establishes connections with the cisterna chyli but loses its connections with neighbouring veins.

The last of the lymph sacs, the paired posterior lymph sacs, develop from the iliac veins. The posterior lymph sacs produce capillary plexuses and lymphatic vessels of the abdominal wall, pelvic region, and lower limbs. The posterior lymph sacs join the cisterna chyli and lose their connections with adjacent veins.

With the exception of the anterior part of the sac from which the cisterna chyli develops, all lymph sacs become invaded by mesenchymal cells and are converted into groups of lymph nodes.

The spleen develops from mesenchymal cells between layers of the dorsal mesentery of the stomach. The thymus arises as an outgrowth of the third pharyngeal pouch.

Function

The lymphatic system has multiple interrelated functions:

- It is responsible for the removal of interstitial fluid from tissues.
- It absorbs and transports fatty acids and fats as chyle from the digestive system.
- It transports white blood cells to and from the lymph nodes into the bones.

- The lymph transports antigen-presenting cells, such as dendritic cells, to the lymph nodes where an immune response is stimulated.

Fat Absorption

Lymph vessels called lacteals are at the beginning of the gastrointestinal tract, predominantly in the small intestine. While most other nutrients absorbed by the small intestine are passed on to the portal venous system to drain via the portal vein into the liver for processing, fats (lipids) are passed on to the lymphatic system to be transported to the blood circulation via the thoracic duct. (There are exceptions, for example medium-chain triglycerides are fatty acid esters of glycerol that passively diffuse from the GI tract to the portal system.) The enriched lymph originating in the lymphatics of the small intestine is called chyle. The nutrients that are released into the circulatory system are processed by the liver, having passed through the systemic circulation.

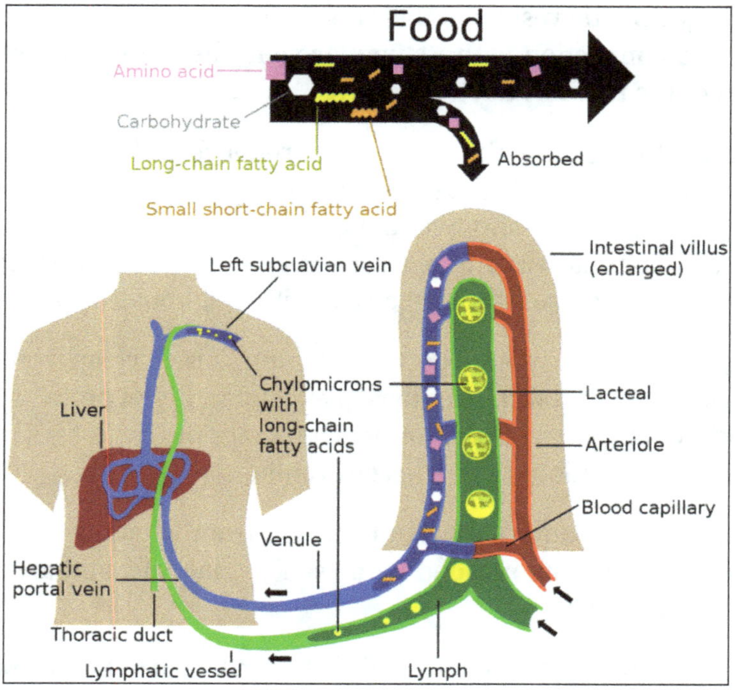

Nutrients in food are absorbed via intestinal vili (greatly enlarged in the picture) to blood and lymph. Long-chain fatty acids (and other lipids with similar fat solubility like some medicines) are absorbed to the lymph and move in it enveloped inside chylomicrons. They move via the thoracic duct of the lymphatic system and finally enter the blood via the left subclavian vein, thus bypassing the liver's first-pass metabolism completely.

Immune Function

The lymphatic system plays a major role in the body's immune system, as the primary site for cells relating to adaptive immune system including T-cells and B-cells. Cells in the lymphatic system react to antigens presented or found by the cells directly or by other dendritic cells. When an antigen is recognized, an immunological cascade begins involving the activation and recruitment of more and more cells, the production of antibodies and cytokines and the recruitment of other immunological cells such as macrophages.

Endocrine System

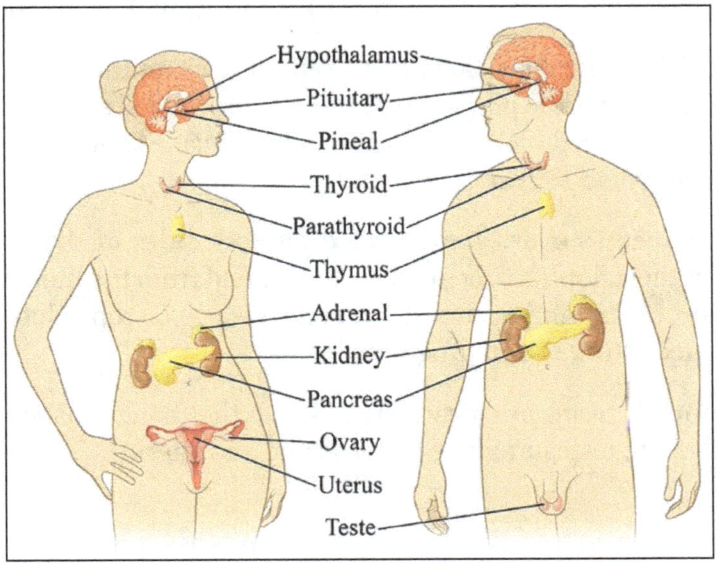

The endocrine system is a complex network of glands and organs. It uses hormones to control and coordinate your body's metabolism energy level, reproduction, growth and development, and response to injury, stress, and mood. The following are integral parts of the endocrine system:

- Hypothalamus: The hypothalamus is located at the base of the brain, near the optic chiasm where the optic nerves behind each eye cross and meet. The hypothalamus secretes hormones that stimulate or suppress the release of hormones in the pituitary gland, in addition to controlling water balance, sleep, temperature, appetite, and blood pressure.
- Pineal body: The pineal body is located below the corpus callosum, in the middle of the brain. It produces the hormone melatonin, which helps the body know when it's time to sleep.
- Pituitary: The pituitary gland is located below the brain. Usually no larger than a pea, the gland controls many functions of the other endocrine glands.

- Thyroid and parathyroid: The thyroid gland and parathyroid glands are located in front of the neck, below the larynx (voice box). The thyroid plays an important role in the body's metabolism. The parathyroid glands play an important role in the regulation of the body's calcium balance.
- Thymus: The thymus is located in the upper part of the chest and produces white blood cells that fight infections and destroy abnormal cells.
- Adrenal gland: An adrenal gland is located on top of each kidney. Like many glands, the adrenal glands work hand-in-hand with the hypothalamus and pituitary gland. The adrenal glands make and release corticosteroid hormones and epinephrine that maintain blood pressure and regulate metabolism.
- Pancreas: The pancreas is located across the back of the abdomen, behind the stomach. The pancreas plays a role in digestion, as well as hormone production. Hormones produced by the pancreas include insulin and glucagon, which regulate levels of blood sugar.
- Ovary: A woman's ovaries are located on both sides of the uterus, below the opening of the fallopian tubes (tubes that extend from the uterus to the ovaries). In addition to containing the egg cells necessary for reproduction, the ovaries also produce estrogen and progesterone.
- Testis: A man's testes are located in a pouch that hangs suspended outside the male body. The testes produce testosterone and sperm.

Function

Hormones

A hormone is any of a class of signaling molecules produced by glands in multicellular organisms that are transported by the circulatory system to target distant organs to regulate physiology and behaviour. Hormones have diverse chemical structures, mainly of 3 classes: eicosanoids, steroids, and amino acid/protein derivatives (amines, peptides, and proteins). The glands that secrete hormones comprise the endocrine system. The term hormone is sometimes extended to include chemicals produced by cells that affect the same cell (autocrine or intracrine signalling) or nearby cells (paracrine signalling).

Hormones are used to communicate between organs and tissues for physiological regulation and behavioral activities, such as digestion, metabolism, respiration, tissue function, sensory perception, sleep, excretion, lactation, stress, growth and development, movement, reproduction, and mood.

Hormones affect distant cells by binding to specific receptor proteins in the target cell resulting in a change in cell function. This may lead to cell type-specific responses that include rapid changes to the activity of existing proteins, or slower changes in the expression of target genes. Amino acid–based hormones (amines and peptide or protein hormones) are water-soluble and act on the surface of target cells via signal transduction pathways; steroid hormones, being lipid-soluble, move through the plasma membranes of target cells to act within their nuclei.

Cell Signalling

The typical mode of cell signalling in the endocrine system is endocrine signaling, that is, using the circulatory system to reach distant target organs. However, there are also other modes, i.e., paracrine, autocrine, and neuroendocrine signaling. Purely neurocrine signaling between neurons, on the other hand, belongs completely to the nervous system.

Autocrine

Autocrine signaling is a form of signaling in which a cell secretes a hormone or chemical messenger (called the autocrine agent) that binds to autocrine receptors on the same cell, leading to changes in the cells.

Paracrine

Some endocrinologists and clinicians include the paracrine system as part of the endocrine system, but there is not consensus. Paracrines are slower acting, targeting cells in the same tissue or organ. An example of this is somatostatin which is released by some pancreatic cells and targets other pancreatic cells.

Juxtacrine

Juxtacrine signaling is a type of intercellular communication that is transmitted via oligosaccharide, lipid, or protein components of a cell membrane, and may affect either the emitting cell or the immediately adjacent cells.

It occurs between adjacent cells that possess broad patches of closely opposed plasma membrane linked by transmembrane channels known as connexons. The gap between the cells can usually be between only 2 and 4 nm.

Excretory System

The excretory system is a passive biological system that removes excess, unnecessary materials from the body fluids of an organism, so as to help maintain internal chemical homeostasis and prevent damage to the body. The dual function of excretory systems is the elimination of the waste products of metabolism and to drain the body of used up and broken down components in a liquid and gaseous state. In humans and other amniotes (mammals, birds and reptiles) most of these substances leave the body as urine and to some degree exhalation, mammals also expel them through sweating.

Only the organs specifically used for the excretion are considered a part of the excretory system. In the narrow sense, the term refers to the urinary system. However, as excretion involves several functions that are only superficially related, it is not usually used in more formal classifications of anatomy or function.

As most healthy functioning organs produce metabolic and other wastes, the entire organism depends on the function of the system. Breaking down of one of more of the systems is a serious health condition, for example kidney failure.

Systems

Urinary System

The kidneys are large, bean-shaped organs which are present on each side of the vertebral column in the abdominal cavity. Humans have two kidneys and each kidney is supplied with blood from the renal artery. The kidneys remove from the blood the nitrogenous wastes such as urea, as well as salts and excess water, and excrete them in the form of urine. This is done with the help of millions of nephrons present in the kidney. The filtrated blood is carried away from the kidneys by the renal vein (or kidney vein). The urine from the kidney is collected by the ureter (or excretory tubes), one from each kidney, and is passed to the urinary bladder. The urinary bladder collects and stores the urine until urination. The urine collected in the bladder is passed into the external environment from the body through an opening called the urethra.

Kidneys

The kidney's primary function is the elimination of waste from the bloodstream by production of urine. They perform several homeostatic functions such as:

- Maintain volume of extracellular fluid.
- Maintain ionic balance in extracellular fluid.
- Maintain pH and osmotic concentration of the extracellular fluid.
- Excrete toxic metabolic by-products such as urea, ammonia, and uric acid.

The way the kidneys do this is with nephrons. There are over 1 million nephrons in each kidney; these nephrons act as filters inside the kidneys. The kidneys filter needed materials and waste, the needed materials go back into the bloodstream, and unneeded materials become urine and are gotten rid of.

In some cases, excess wastes crystallize as kidney stones. They grow and can become painful irritants that may require surgery or ultrasound treatments. Some stones are small enough to be forced into the urethra.

Ureter

The ureters are muscular ducts that propel urine from the kidneys to the urinary bladder. In the human adult, the ureters are usually 25–30 cm (10–12 in) long. In humans, the ureters arise from the renal pelvis on the medial aspect of each kidney before descending towards the bladder on the front of the psoas major muscle. The ureters cross the pelvic brim near the bifurcation of the iliac arteries (which they run over). This "pelviureteric junction" is a common site for the impaction of kidney stones (the other being the uteterovesical valve). The ureters run posteriorly on the lateral walls of the pelvis. They then curve anteriormedially to enter the bladder through the back, at the vesicoureteric junction, running within the wall of the bladder for a few centimeters. The backflow of urine is prevented by valves known as ureterovesical valves. In the female, the ureters pass through the mesometrium on the way to the bladder.

Urinary Bladder

The urinary bladder is the organ that collects waste excreted by the kidneys prior to disposal by urination. It is a hollow muscular, and distensible (or elastic) organ, and sits on the pelvic floor. Urine enters the bladder via the ureters and exits via the urethra.

Embryologically, the bladder is derived from the urogenital sinus, and it is initially continuous with the allantois. In human males, the base of the bladder lies between the rectum and the pubic symphysis. It is superior to the prostate, and separated from the rectum by the rectovesical excavation. In females, the bladder sits inferior to the uterus and anterior to the vagina. It is separated from the uterus by the vesicouterine excavation. In infants and young children, the urinary bladder is in the abdomen even when empty.

Urethra

In anatomy, it is a tube which connects the urinary bladder to the outside of the body. In humans, the urethra has an excretory function in both genders to pass.

Respiratory System

One of the main functions of the lungs is to diffuse gaseous wastes, such as carbon dioxide, from the bloodstream as a normal part of respiration.

Gastrointestinal Tract

The large intestine's main function is to transport food particles through the body and expel the indigestible parts at the other end, but it also collects waste from throughout the body. The typical brown colour of mammal waste is due to bilirubin, a breakdown product of normal heme catabolism. The lower part of the large intestine also extracts any remaining usable water and then removes solid waste. At about 10 feet long in humans, it transports the wastes through the tubes to be excreted.

Biliary System

The liver detoxifies and breaks down chemicals, poisons and other toxins that enter the body. For example, the liver transforms ammonia (which is poisonous) into urea in fish, amphibians and mammals, and into uric acid in birds and reptiles. Urea is filtered by the kidney into urine or through the gills in fish and tadpoles. Uric acid is paste-like and expelled as a semi-solid waste (the "white" in bird excrements). The liver also produces bile, and the body uses bile to break down fats into usable fats and unusable waste.

Invertebrates lack a liver, but most terrestrial groups, like insects, possesses a number of blind guts that serve the similar functions. Marine invertebrates do not need the ammonia conversion of the liver, as they can usually expel ammonia directly by diffusion through the skin.

Integumentary System

Skin

Sweat glands in the skin secrete a fluid waste called sweat or perspiration; however, its primary functions are temperature control and pheromone release. Therefore, its role as a part of the excretory system is minimal. Sweating also maintains the level of salt in the body.

In mammals, the skin excretes sweat through sweat glands throughout the body. The sweat, helped by salt, evaporates and helps to keep the body cool when it is warm. In amphibians, the lungs are very simple, and they lack the necessary means to the exhale like other tetrapods can. The moist, scale-less skin is therefore essential in helping to rid the blood of carbon dioxide, and also allows for urea to be expelled through diffusion when submerged.

In small-bodied marine invertebrates, the skin is the most important excretory organ. That is particularly true for acoelomate groups like cnidarians, flatworms and nemerteans, who have no body cavities and hence no body fluid that can be drained or purified by nephrons, which is the reason acoelomate animals are thread-like (nemertans), flat (flatworms) or only consist of a thin layer of cells around a gelatinous non-cellular interior (cnidarians).

Eccrine

Like sweat glands, eccrine glands allow excess water to leave the body. The majority of eccrine glands are located mainly on the forehead, the bottoms of the feet, and the palms, although the glands are everywhere throughout the body. They help the body to maintain temperature control. Eccrine glands in the skin are unique to mammals.

Secretions of sweat from the eccrine glands play a large role in controlling the body temperature of humans. Regulation of body temperature, also known as thermoregulation, is very important when it comes to instances that bring the body's temperature outside of the homeostatic temperature such as with a fever or even exercise. Together these glands make up the size of about one kidney and in one day a human can perspire amounts as much as 10 liters. The two functions consist of secretion of a filtrate in response to acetylcholine and reabsorption of sodium near the duct when there is water in excess so that a sweat can be surfacing the skin.

There are three parts to the eccrine sweat gland and these are the pore, the duct, and the gland. The pore is the portion that goes through the outermost layer of the skin and is typically 5-10 microns in diameter. The duct is the part of the sweat gland that connects dermis cells to the epidermis. It is composed by two layers of cells and is between 10 and 20 microns in diameter. The gland does the actual secretion and it lies deep within the dermis. The cells that make up the gland are larger in size than the duct cells and its lumen is around 20 microns in diameter.

Substances

Bile

After bile is produced in the liver, it is stored in the gall bladder. It is then secreted within the small intestine where it helps to emulsify fats in the same manner as a soap. Bile also contains bilirubin, which is a waste product.

Bile salts can be considered waste that is useful for the body given that they have a role in fat absorption from the stomach. They are excreted from the liver and along with blood flow they help to form the shape of the liver where they are excreted. For instance, if biliary drainage is impaired than that part of the liver will end up wasting away.

Biliary obstruction is typically due to masses blocking the ducts of the system such as tumors. The consequences of this depend on the site of blockage and how long it goes on for. There is inflammation of the ducts due to the irritation from the bile acids and this can cause infections. If rupture of the duct takes place it is very traumatic and even fatal.

Urine

Within the kidney, blood first passes through the afferent artery to the capillary formation called a glomerulus and is collected in the Bowman's capsule, which filters the blood from its contents—primarily food and wastes. After the filtration process, the blood then returns to collect the food nutrients it needs, while the wastes pass into the collecting duct, to the renal pelvis, and to the ureter, and are then secreted out of the body via the urinary bladder.

Digestive System

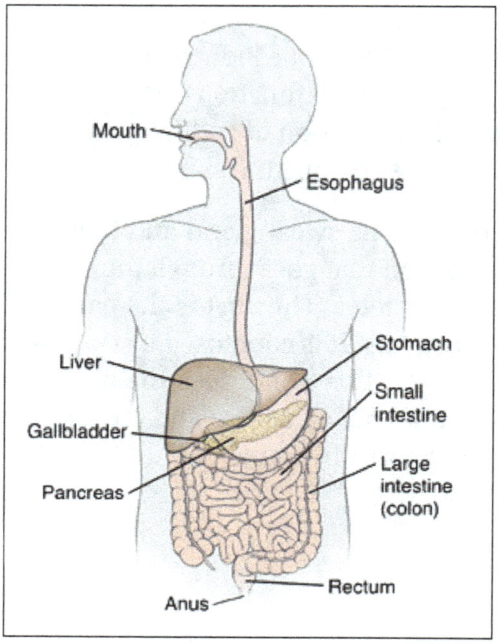

The digestive system breaks food down into basic nutrients that can be used by the body. The digestive tract is a long, muscular tube that extends from the mouth through the stomach and intestines to the anus. As food moves along the digestive tract, it is digested. This means it is changed into substances that can be absorbed into the bloodstream. Certain organs (such as the liver, gallbladder, and pancreas) help with this digestion. Parts of food that can't be digested are turned into stool. This is waste material that is passed out of the body.

The digestive system is made up of the following:

- The mouth: Takes in food, breaks it into pieces, and begins the digestion process.
- The esophagus: Moves food from the mouth to the stomach.
- The stomach: Breaks food down into a liquid mixture.
- The liver: Makes bile that helps digest fat.
- The gallbladder: Stores bile.
- The pancreas: Makes enzymes that help in digestion.

- The small intestine: Digests food further and absorbs nutrients. What is left is passed on to the colon as liquid waste.
- The large intestine (colon): Absorbs water, salt, and minerals from the waste, forming a solid stool.
- The rectum: Stores stool until a bowel movement happens.
- The anus: The opening where stool leaves the body.

Urinary System

The urinary system's function is to filter blood and create urine as a waste by-product. The organs of the urinary system include the kidneys, renal pelvis, ureters, bladder and urethra.

The body takes nutrients from food and converts them to energy. After the body has taken the food components that it needs, waste products are left behind in the bowel and in the blood.

The kidney and urinary systems help the body to eliminate liquid waste called urea, and to keep chemicals, such as potassium and sodium, and water in balance. Urea is produced when foods containing protein, such as meat, poultry, and certain vegetables, are broken down in the body. Urea is carried in the bloodstream to the kidneys, where it is removed along with water and other wastes in the form of urine.

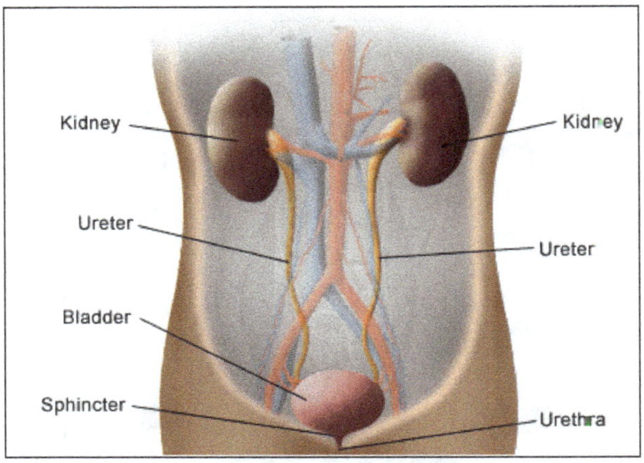

Front View of Urinary Tract.

Other important functions of the kidneys include blood pressure regulation and the production of erythropoietin, which controls red blood cell production in the bone marrow. Kidneys also regulate the acid-base balance and conserve fluids.

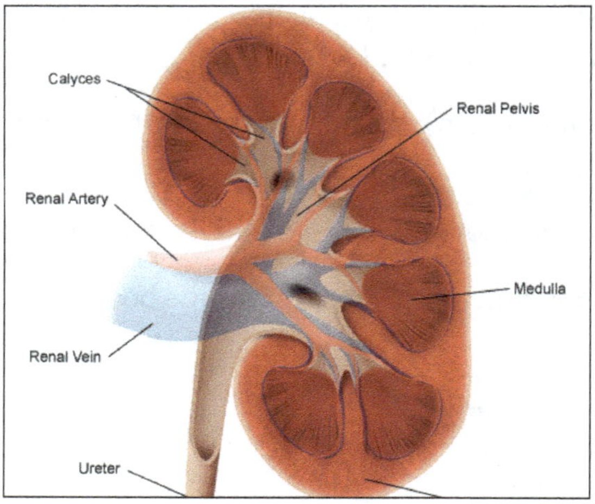

Anatomy of the Kidney.

Kidney and Urinary System Parts and their Functions

- Two kidneys: This pair of purplish-brown organs is located below the ribs toward the middle of the back. Their function is to:
  - Remove waste products and drugs from the body.
  - Balance the body's fluids.
  - Release hormones to regulate blood pressure.
  - Control production of red blood cells.
  - Diagram of the bladder.

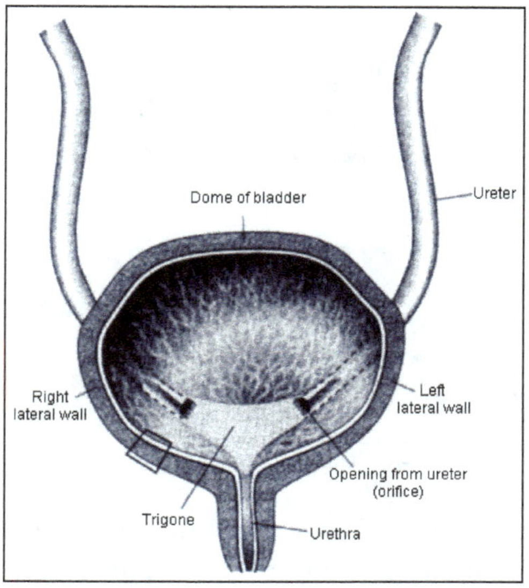

The kidneys remove urea from the blood through tiny filtering units called nephrons.

Each nephron consists of a ball formed of small blood capillaries, called a glomerulus, and a small tube called a renal tubule. Urea, together with water and other waste substances, forms the urine as it passes through the nephrons and down the renal tubules of the kidney.

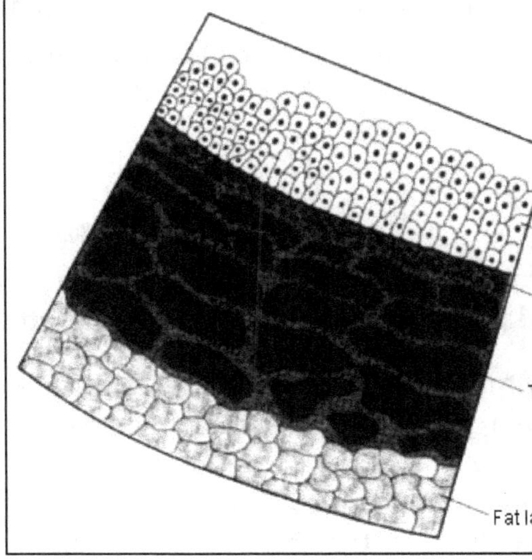

- Two ureters: These narrow tubes carry urine from the kidneys to the bladder. Muscles in the ureter walls continually tighten and relax forcing urine downward, away from the kidneys. If urine backs up, or is allowed to stand still, a kidney infection can develop. About every 10 to 15 seconds, small amounts of urine are emptied into the bladder from the ureters.
- Bladder: This triangle-shaped, hollow organ is located in the lower abdomen. It is held in place by ligaments that are attached to other organs and the pelvic bones. The bladder's walls relax and expand to store urine, and contract and flatten to empty urine through the urethra. The typical healthy adult bladder can store up to two cups of urine for two to five hours. Upon examination, specific "landmarks" are used to describe the location of any irregularities in the bladder. These are:
    - Trigone: A triangle-shaped region near the junction of the urethra and the bladder.
    - Right and left lateral walls: Walls on either side of the trigone.
    - Posterior wall: Back wall.
    - Dome: Roof of the bladder.
- Two sphincter muscles: These circular muscles help keep urine from leaking by closing tightly like a rubber band around the opening of the bladder.

- Nerves in the bladder: The nerves alert a person when it is time to urinate, or empty the bladder.
- Urethra: This tube allows urine to pass outside the body. The brain signals the bladder muscles to tighten, which squeezes urine out of the bladder. At the same time, the brain signals the sphincter muscles to relax to let urine exit the bladder through the urethra. When all the signals occur in the correct order, normal urination occurs.

Structure

The urinary system refers to the structures that produce and transport urine to the point of excretion. In the human urinary system there are two kidneys that are located between the dorsal body wall and parietal peritoneum on both the left and right sides.

The formation of urine begins within the functional unit of the kidney, the nephrons. Urine then flows through the nephrons, through a system of converging tubules called collecting ducts. These collecting ducts then join together to form the minor calyces, followed by the major calyces that ultimately join the renal pelvis. From here, urine continues its flow from the renal pelvis into the ureter, transporting urine into the urinary bladder. The anatomy of the human urinary system differs between males and females at the level of the urinary bladder. In males, the urethra begins at the internal urethral orifice in the trigone of the bladder, continues through the external urethral orifice, and then becomes the prostatic, membranous, bulbar, and penile urethra. Urine exits through the external urethral meatus. The female urethra is much shorter, beginning at the bladder neck and terminating in the vaginal vestibule.

Microanatomy

Under microscopy, the urinary system is covered in a unique lining called urothelium, a type of transitional epithelium. Unlike the epithelial lining of most organs, transitional epithelium can flatten and distend. Urothelium covers most of the urinary system, including the renal pelvis, ureters, and bladder.

Function

The main functions of the urinary system and its components are to:

- Regulate blood volume and composition (e.g. sodium, potassium and calcium).
- Regulate blood pressure.
- Regulate pH homeostasis of the blood.
- Contributes to the production of red blood cells by the kidney.
- Helps synthesize calcitrol the (active form of Vitamin D).

- Stores waste product (mainly urea and uric acid) before it and other products are removed from the body.

Urine Formation

Average urine production in adult humans is about 1–2 litres (L) per day, depending on state of hydration, activity level, environmental factors, weight, and the individual's health. Producing too much or too little urine requires medical attention. Polyuria is a condition of excessive urine production (> 2.5 L/day). Oliguria when < 400 mL (millilitres) are produced, and anuria one of < 100 mL per day.

The first step in urine formation is the filtration of blood in the kidneys. In a healthy human the kidney receives between 12 and 30% of cardiac output, but it averages about 20% or about 1.25 L/min.

The basic structural and functional unit of the kidney is the nephron. Its chief function is to regulate the concentration of water and soluble substances like sodium by filtering the blood, reabsorbing what is needed and excreting the rest as urine.

In the first part of the nephron, Bowman's capsule filters blood from the circulatory system into the tubules. Hydrostatic and osmotic pressure gradients facilitate filtration across a semipermeable membrane. The filtrate includes water, small molecules, and ions that easily pass through the filtration membrane. However larger molecules such as proteins and blood cells are prevented from passing through the filtration membrane. The amount of filtrate produced every minute is called the glomerular filtration rate or GFR and amounts to 180 litres per day. About 99% of this filtrate is reabsorbed as it passes through the nephron and the remaining 1% becomes urine.

The urinary system is regulated by the endocrine system by hormones such as antidiuretic hormone, aldosterone, and parathyroid hormone.

Regulation of Concentration and Volume

The urinary system is under influence of the circulatory system, nervous system, and endocrine system.

Aldosterone plays a central role in regulating blood pressure through its effects on the kidney. It acts on the distal tubules and collecting ducts of the nephron and increases reabsorption of sodium from the glomerular filtrate. Reabsorption of sodium results in retention of water, which increases blood pressure and blood volume. Antidiuretic hormone (ADH), is a neurohypophysial hormone found in most mammals. Its two primary functions are to retain water in the body and vasoconstriction. Vasopressin regulates the body's retention of water by increasing water reabsorption in the collecting ducts of the kidney nephron. Vasopressin increases water permeability of the kidney's collecting duct and distal convoluted tubule by inducing translocation of aquaporin-CD water channels in the kidney nephron collecting duct plasma membrane.

Urination

Urination is the ejection of urine from the urinary bladder through the urethra to the outside of the body. In healthy humans (and many other animals), the process of urination is under voluntary control. In infants, some elderly individuals, and those with neurological injury, urination may occur as an involuntary reflex. Physiologically, micturition involves coordination between the central, autonomic, and somatic nervous systems. Brain centers that regulate urination include the pontine micturition center, periaqueductal gray, and the cerebral cortex. In placental mammals the male ejects urine through the penis, and the female through the vulva.

Integumentary System

The skin and its derivatives (sweat and oil glands, hair and nails) serve a number of functions, mostly protective; together, these organs are called the integumentary system.

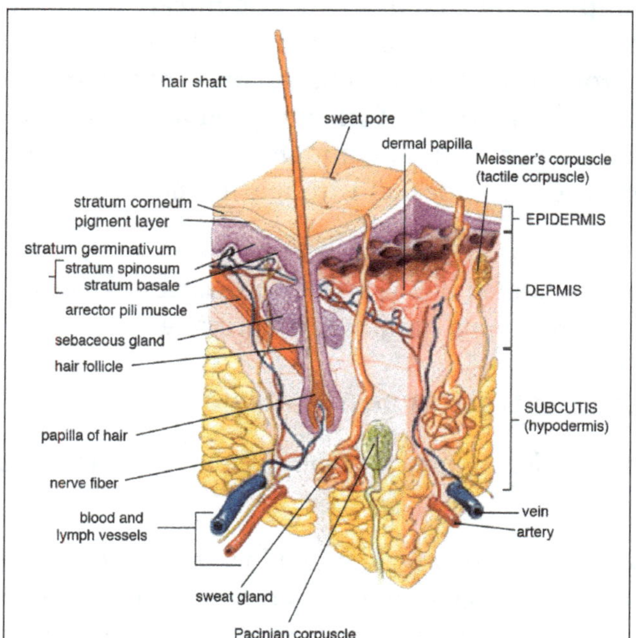

Integumentary System.

Structure of the Skin

The skin is composed of two kinds of tissue: The outer epidermis and the underlying dermis.

Epidermis

The outer epidermis composed of stratified squamous epithelium that is capable of keratinizing or becoming hard and tough.

- Composition: The epidermis is composed of up to five layers or strata; from the inside out these are the: stratum basale, spinosum, granulosum, lucidum, and corneum.
- Epithelial tissue: Like all other epithelial tissues, the epidermis is avascular; that is, it has no blood supply of its own.
- Keratinocytes: Most cells of the epidermis are keratinocytes (keratin cells), which produce keratin, the fibrous protein that makes the epidermis a tough protective layer.
- Stratum basale: The deepest layer of the epidermis, the stratum basale, lies closest to the dermis and is connected to it along a wavy a borderline that resembles corrugated cardboard; this basal layer contains epidermal cells that receive the most adequate nourishment via diffusion of nutrients from the dermis.
- Stratum spinosum: As the epidermal layers move away from the dermis and become part of the more superficial layers, the stratum spinosum.
- Stratum granulosum: Upon reaching the stratum granulosum, the layers become flatter and increasingly full of keratin.
- Stratum lucidum: Finally, they die, forming the clear stratum lucidum; this latter epidermal layer is not present in all skin regions, it occurs only where the skin is hairless and extra thick, that is, on the palms of the hands and soles of the feet.
- Stratum corneum: The outermost layer, the stratum corneum, is 20 to 30 cells layers thick but it accounts for about three-quarters of epidermal thickness; it rubs and flakes off slowly and steadily as the dandruff familiar to everyone; then, this layer is replaced by cells produced by the division of the deeper stratum basale cells.
- Cornified cells: The shinglelike dead cell remnants, completely filled with keratin, are referred to as cornified or horny cells.
- Keratin: Keratin is an exceptionally tough protein; its abundance in the stratum corneum allows that layer to provide a durable "overcoat" for the body, which protects deeper cells from the hostile external environment.
- Melanin: Melanin, a pigment that ranges in color from yellow to brown to black,

is produced by special spider-shaped cells called melanocytes, found chiefly in the stratum basale.

- Melanosomes: As the melanocytes produce melanin, it accumulates within them in membrane-bound granules called melanosomes; these granules then move to the ends of the spidery arms of the melanocytes, where they are taken up by nearby keratinocytes.

Dermis

The underlying dermis is mostly made up of dense connective tissue.

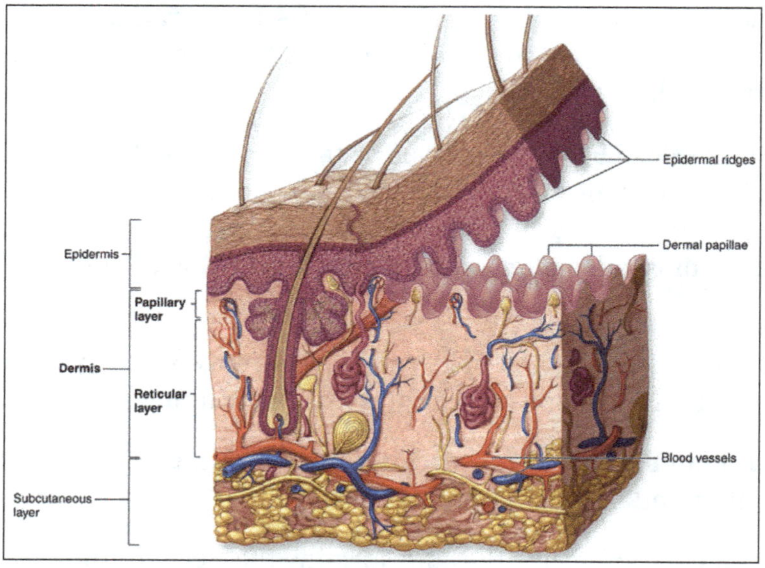

Layers of the skin.

- Major regions: The dense (fibrous) connective tissue making up the dermis consists of two major regions- the papillary and reticular regions.
- Papillary layer: The papillary layer is the upper dermal region; it is uneven and has peglike projections from its superior surface called dermal papillae, which indent the epidermis above and contain capillary loops which furnish nutrients to the epidermis; it also has papillary patterns that form looped and whorled ridges on the epidermal surface that increase friction and enhance the gripping ability of the fingers and feet.
- Reticular layer: The reticular layer is the deepest skin layer; it contains blood vessels, sweat and oil glands, and deep pressure receptors called Pacinian corpuscles.
- Collagen: Collagen fibers are responsible for the toughness of the dermis; they also attract and bind water and thus help to keep the skin hydrated.

- Elastic fibers: Elastic fibers give the skin its elasticity when we are young, and as we age, the number of collagen and elastic fibers decreases and the subcutaneous tissue loses fat.
- Blood vessels: The dermis is abundantly supplied with blood vessels that play a role in maintaining body temperature homeostasis; when body temperature is high, the capillaries of the dermis becomes engorged, or swollen, with heated blood, and the skin becomes reddened and warm; if the environment is cool, blood bypasses the dermis capillaries temporarily, allowing internal body temperature to stay high.
- Nerve supply: The dermis also has a rich nerve supply; many of the nerve endings have specialized receptor end-organs that send messages to the central nervous system for interpretation when they are stimulated by environmental factors.

Appendages of the Skin

The skin appendages include cutaneous glands, hair and hair follicle, and nails.

Cutaneous Glands

As these glands are formed by the cells of the stratum basale, they push into deeper skin regions and ultimately reside almost entirely in the dermis:

- Exocrine glands: The cutaneous glands are all exocrine glands that release their secretions to the skin surface via ducts and they fall into two groups: sebaceous glands and sweat glands.
- Sebaceous (oil) glands: The sebaceous, or oil, glands are found all over the skin, except on the palms of the hands and the soles of the feet; their ducts usually empty into a hair follicle; the product of the sebaceous glands, sebum, is a mixture of oily substances and fragmented cells, and it is a lubricant that keeps the skin soft and moist and prevents the hair from becoming brittle.
- Sweat glands: Sweat glands, also called sudoriferous glands, are widely distributed in the skin, and there are two types: eccrine and apocrine.
- Eccrine glands: The eccrine glands are far more numerous and are found all over the body; they produce sweat, a clear secretion that is primarily water plus some salts, vitamin C, trace of metabolic wastes, and lactic acid; the eccrine glands are also a part of the body's heat regulating equipment.
- Apocrine glands: Apocrine glands are largely confined to the axillary and genital areas of the body; they are usually larger than eccrine glands and their ducts empty into hair follicles; their secretion contain fatty acids and proteins, as well

as all substances present in eccrine secretion; they begin to function during puberty under the influence of androgens, and they also play a minimal role in thermoregulation.

Hair and Hair Follicles

There are millions of hair scattered all over the body, but other than serving a few minor protective functions, our body hair has lost much of its usefulness.

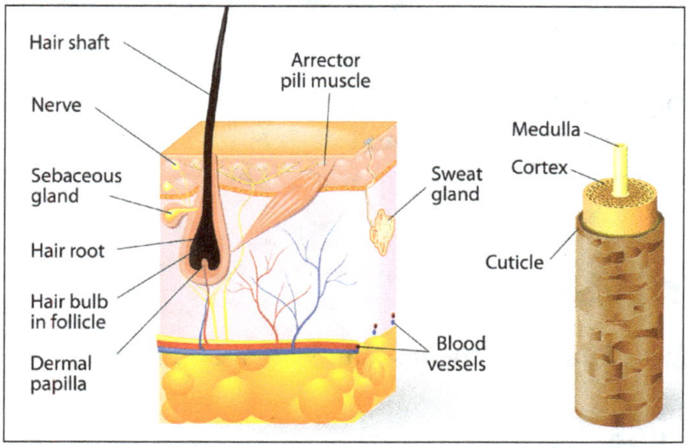

Hair anatomy.

- Hairs: A hair, produced by a hair follicle, is a flexible epithelial structure.
- Root: The part of the hair enclosed in the follicle is the root.
- Shaft: The part projecting from the surface of the scalp or skin is called shaft.
- Formation: The hair is formed by division of a well-nourished stratum basale epithelial cells in the matrix (growth zone) of the hair bulb at the inferior end of the follicle.
- Composition: Each hair is made up of a central core called the medulla surrounded by a bulky cortex layer.
- Cuticle: The cortex is enclosed by an outermost cuticle formed by a single layer of cells that overlap one another like shingles on the roof; this arrangement helps to keep the hairs apart and keeps them from matting; the cuticle is the most heavily keratinized region; it provides strength and helps keep the inner hair layers tightly compacted.
- Hair pigment: Hair pigment is made by melanocytes in the hair bulb, and varying amounts of different types of melanin combine to produce all varieties of hair color from pale blond to pitch black.
- Hair follicles: Hair follicles are actually compound structures.

- Epidermal sheath: The inner epidermal sheath is composed of epithelial tissue and forms the hair.
- Dermal sheath: The outer dermal sheath is actually dermal connective tissue; this dermal region supplies blood vessels to the epidermal portion and reinforces it.
- Papilla: Its nipple-like papilla provides the blood supply to the matrix in the hair bulb.
- Arrector pili: Small bands of smooth muscle cells-arrector pili-connect each side of the hair follicle to the dermal tissue; when these muscles contract, the hair is pulled upright, dimpling the skin surface with “goosebumps”.

Nails

A nail is a scalelike modification of the epidermis that corresponds to the hoof or claw of other animals.

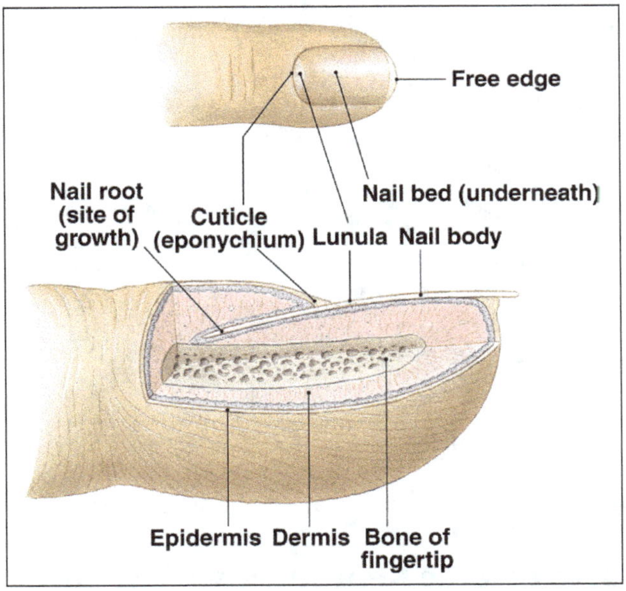

Nail Anatomy.

- Parts: Each nail has a free edge, a body (visible attached portion), and a root (embedded in the skin).
- Nail folds: The borders of the nail are overlapped by skin folds, called nail folds.
- Cuticle: The thick proximal nail fold is commonly called the cuticle.
- Nail bed: The stratum basale of the epidermis extends beneath the nail as the nail bed.

- Nail matrix: Its thickened proximal area, the nail matrix, is responsible for nail growth.
- Color: Nails are transparent and nearly colorless, but they look pink because of the rich blood supply in the underlying dermis.
- Lunula: The exception to the pinkish color of the nails is the region over the thickened nail matrix that appears as a white crescent and is called the lunula.

Male Reproductive System

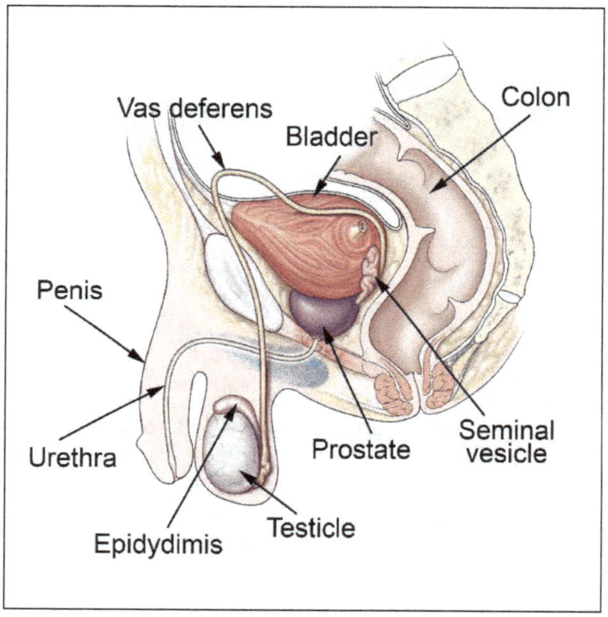

Male Reproductive Anatomy.

The male reproductive anatomy includes internal and external structures.

External Structures of Male Reproductive System

Most of the male reproductive system is located outside of the man's abdominal cavity or pelvis. The external structures of the male reproductive system are the penis, the scrotum and the testicles.

- Penis: The penis is the male organ for sexual intercourse. It has three parts: the root, which attaches to the wall of the abdomen; the body, or shaft; and the glans, which is the cone-shaped end of the penis. The glans, which also is called the head of the penis, is covered with a loose layer of skin called foreskin. (This skin is sometimes removed in a procedure called circumcision.) The opening of the urethra, the tube that transports semen and urine, is at the tip of the glans penis. The penis also contains a number of sensitive nerve endings. The body of the penis is cylindrical in shape and consists of three internal chambers. These chambers are made up of special, sponge-like erectile tissue. This tissue contains thousands of large spaces that fill with blood when the man is sexually aroused. As the penis fills with blood, it becomes rigid and erect, which allows for penetration during sexual intercourse. The skin of the penis is loose and elastic to allow for changes in penis size during an erection.

- Semen, which contains sperm, is expelled (ejaculated) through the end of the penis when the man reaches sexual climax (orgasm). When the penis is erect, the flow of urine is blocked from the urethra, allowing only semen to be ejaculated at orgasm.
- Scrotum: The scrotum is the loose pouch-like sac of skin that hangs behind the penis. It contains the testicles (also called testes), as well as many nerves and blood vessels. The scrotum has a protective function and acts as a climate control system for the testes. For normal sperm development, the testes must be at a temperature slightly cooler than the body temperature. Special muscles in the wall of the scrotum allow it to contract (tighten) and relax, moving the testicles closer to the body for warmth and protection or farther away from the body to cool the temperature.
- Testicles (testes): The testes are oval organs about the size of very large olives that lie in the scrotum, secured at either end by a structure called the spermatic cord. Most men have two testes. The testes are responsible for making testosterone, the primary male sex hormone, and for producing sperm. Within the testes are coiled masses of tubes called seminiferous tubules. These tubules are responsible for producing the sperm cells through a process called spermatogenesis.
- Epididymis: The epididymis is a long, coiled tube that rests on the backside of each testicle. It functions in the carrying and storage of the sperm cells that are produced in the testes. It also is the job of the epididymis to bring the sperm to maturity, since the sperm that emerge from the testes are immature and incapable of fertilization. During sexual arousal, contractions force the sperm into the vas deferens.

Internal Structures of Male Reproductive System

The internal organs of the male reproductive system, also called accessory organs, include the following:

- Vas deferens: The vas deferens is a long, muscular tube that travels from the epididymis into the pelvic cavity, to just behind the bladder. The vas deferens transports mature sperm to the urethra in preparation for ejaculation.

- Ejaculatory ducts: These are formed by the fusion of the vas deferens and the seminal vesicles. The ejaculatory ducts empty into the urethra.
- Urethra: The urethra is the tube that carries urine from the bladder to outside of the body. In males, it has the additional function of expelling (ejaculating) semen when the man reaches orgasm. When the penis is erect during sex, the flow of urine is blocked from the urethra, allowing only semen to be ejaculated at orgasm.
- Seminal vesicles: The seminal vesicles are sac-like pouches that attach to the vas deferens near the base of the bladder. The seminal vesicles produce a sugar-rich fluid (fructose) that provides sperm with a source of energy and helps with the sperms' motility (ability to move). The fluid of the seminal vesicles makes up most of the volume of a man's ejaculatory fluid, or ejaculate.
- Prostate gland: The prostate gland is a walnut-sized structure that is located below the urinary bladder in front of the rectum. The prostate gland contributes additional fluid to the ejaculate. Prostate fluids also help to nourish the sperm. The urethra, which carries the ejaculate to be expelled during orgasm, runs through the center of the prostate gland.
- Bulbourethral glands: The bulbourethral glands, or Cowper's glands, are pea-sized structures located on the sides of the urethra just below the prostate gland. These glands produce a clear, slippery fluid that empties directly into the urethra. This fluid serves to lubricate the urethra and to neutralize any acidity that may be present due to residual drops of urine in the urethra.

Female Reproductive System

- Ovaries: The ovaries are a pair of small glands about the size and shape of almonds, located on the left and right sides of the pelvic body cavity lateral to the superior portion of the uterus. Ovaries produce female sex hormones such as estrogen and progesterone as well as ova (commonly called "eggs"), the female gametes. Ova are produced from oocyte cells that slowly develop throughout a woman's early life and reach maturity after puberty. Each month during ovulation, a mature ovum is released. The ovum travels from the ovary to the fallopian tube, where it may be fertilized before reaching the uterus.
- Fallopian Tubes: The fallopian tubes are a pair of muscular tubes that extend from the left and right superior corners of the uterus to the edge of the ovaries.

The fallopian tubes end in a funnel-shaped structure called the infundibulum, which is covered with small finger-like projections called fimbriae. The fimbriae swipe over the outside of the ovaries to pick up released ova and carry them into the infundibulum for transport to the uterus. The inside of each fallopian tube is covered in cilia that work with the smooth muscle of the tube to carry the ovum to the uterus.

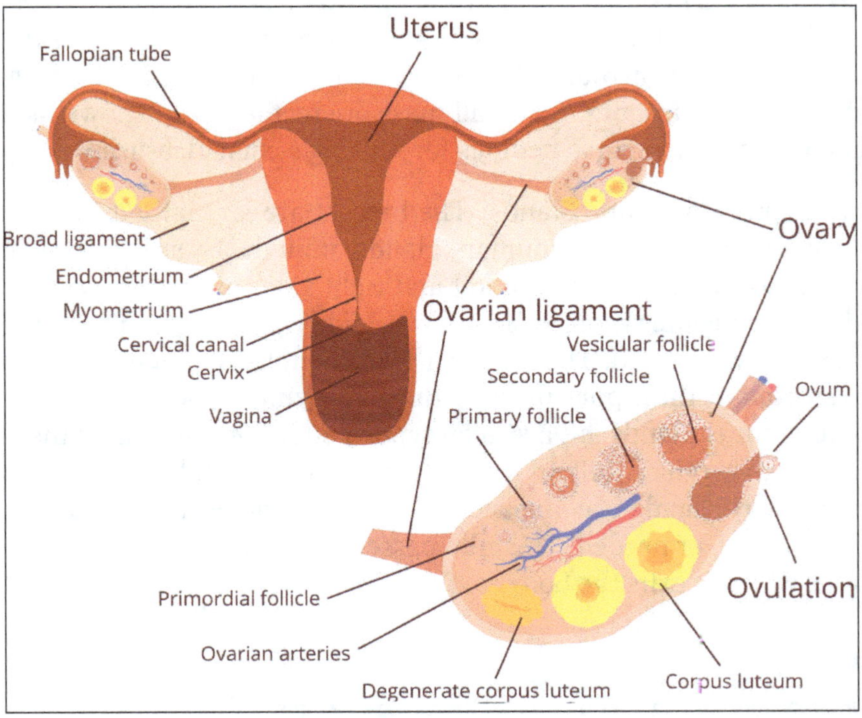

Female Reproductive System.

- Uterus: The uterus is a hollow, muscular, pear-shaped organ located posterior and superior to the urinary bladder. Connected to the two fallopian tubes on its superior end and to the vagina (via the cervix) on its inferior end, the uterus is also known as the womb, as it surrounds and supports the developing fetus during pregnancy. The inner lining of the uterus, known as the endometrium, provides support to the embryo during early development. The visceral muscles of the uterus contract during childbirth to push the fetus through the birth canal.

- Vagina: The vagina is an elastic, muscular tube that connects the cervix of the uterus to the exterior of the body. It is located inferior to the uterus and posterior to the urinary bladder. The vagina functions as the receptacle for the penis during sexual intercourse and carries sperm to the uterus and fallopian tubes. It also serves as the birth canal by stretching to allow delivery of the fetus during childbirth. During menstruation, the menstrual flow exits the body via the vagina.

- Vulva: The vulva is the collective name for the external female genitalia located in the pubic region of the body. The vulva surrounds the external ends of the urethral opening and the vagina and includes the mons pubis, labia majora, labia minora, and clitoris. The mons pubis, or pubic mound, is a raised layer of adipose tissue between the skin and the pubic bone that provides cushioning to the vulva. The inferior portion of the mons pubis splits into left and right halves called the labia majora. The mons pubis and labia majora are covered with pubic hairs. Inside of the labia majora are smaller, hairless folds of skin called the labia minora that surround the vaginal and urethral openings. On the superior end of the labia minora is a small mass of erectile tissue known as the clitoris that contains many nerve endings for sensing sexual pleasure.
- Breasts and Mammary Glands: The breasts are specialized organs of the female body that contain mammary glands, milk ducts, and adipose tissue. The two breasts are located on the left and right sides of the thoracic region of the body. In the center of each breast is a highly pigmented nipple that releases milk when stimulated. The areola, a thickened, highly pigmented band of skin that surrounds the nipple, protects the underlying tissues during breastfeeding. The mammary glands are a special type of sudoriferous glands that have been modified to produce milk to feed infants. Within each breast, 15 to 20 clusters of mammary glands become active during pregnancy and remain active until milk is no longer needed. The milk passes through milk ducts on its way to the nipple, where it exits the body.

References

- Skeletal-system: biologydictionary.net, Retrieved 17 August, 2020
- Allen NJ, Barres BA (2009). "Neuroscience: Glia – more than just brain glue". Nature. 457 (7230): 675–677. doi:10.1038/457675a. PMID 19194443
- Anatomy-digestive-system, health-library: saintlukeskc.org, Retrieved 23 July, 2020
- Blumenfeld, Hal. (2010). Neuroanatomy through clinical cases (2nd ed.). Sunderland, Mass.: Sinauer Associates. ISBN 9780878930586. OCLC 473478856
- Anatomy-of-the-urinary-system, wellness-and-prevention, health: hopkinsmedicine.org, Retrieved 06 April, 2020
- Natale G, Bocci G, Ribatti D (September 2017). "Scholars and scientists in the history of the lymphatic system". Journal of Anatomy. 231 (3): 417–429. doi:10.1111/joa.12644. PMC 5554832. PMID 28614587
- Anatomy-of-the-Integumentary-System, integumentary-system: nurseslabs.com, Retrieved 19 March, 2020

3

Human Head and Neck

The anatomy of the head and neck of human body includes the brain, bones, muscles, blood vessels, nerves, glands, nose, mouth, teeth, tongue, and throat. This chapter closely examines the human head and neck anatomy to provide an easy understanding of the subject.

Head

In human anatomy, head is the upper portion of the body, consisting of the skull with its coverings and contents, including the lower jaw. It is attached to the spinal column by way of the first cervical vertebra, the atlas, and connected with the trunk of the body by the muscles, blood vessels, and nerves that constitute the neck. The term also is used to describe the anterior or fore part of animals other than humans.

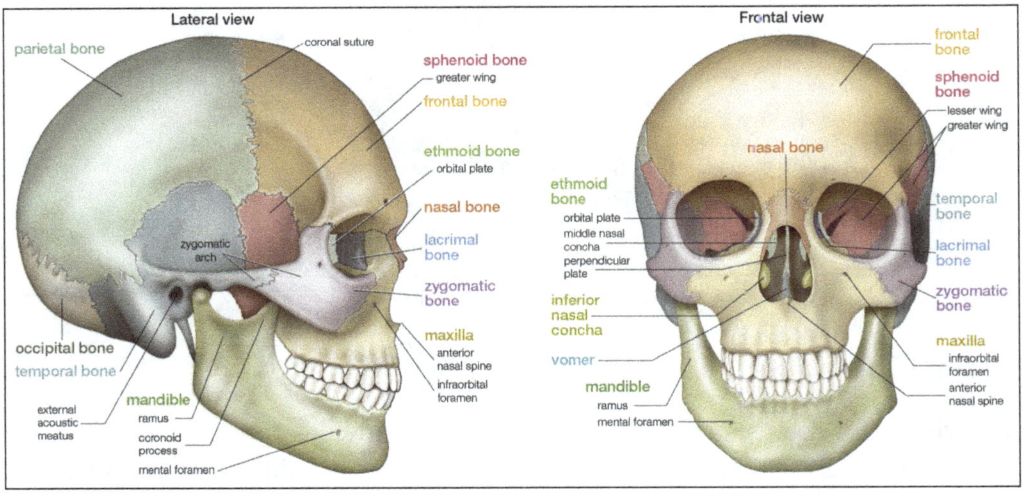

(Left) Lateral and (right) frontal views of the human skull.

Skull is the skeletal framework of the head, composed of bones or cartilage, which form a unit that protects the brain and some sense organs. The upper jaw, but not the lower, is part of the skull. The human cranium, the part that contains the brain, is globular and relatively large in comparison with the face. In most other animals the facial portion of the skull, including the upper teeth and the nose, is larger than the cranium. In humans the skull is supported by the highest vertebra, called the atlas, permitting nodding motion. The atlas turns on the next-lower vertebra, the axis, to allow for side-to-side motion.

In humans the base of the cranium is the occipital bone, which has a central opening (foramen magnum) to admit the spinal cord. The parietal and temporal bones form the sides and uppermost portion of the dome of the cranium, and the frontal bone forms the forehead; the cranial floor consists of the sphenoid and ethmoid bones. The facial area includes the zygomatic, or malar, bones (cheekbones), which join with the temporal and maxillary bones to form the zygomatic arch below the eye socket; the palatine bone; and the maxillary, or upper jaw, bones. The nasal cavity is formed by the vomer and the nasal, lachrymal, and turbinate bones. In infants the sutures (joints) between the various skull elements are loose, but with age they fuse together. Many mammals, such as the dog, have a sagittal crest down the centre of the skull; this provides an extra attachment site for the temporal muscles, which close the jaws.

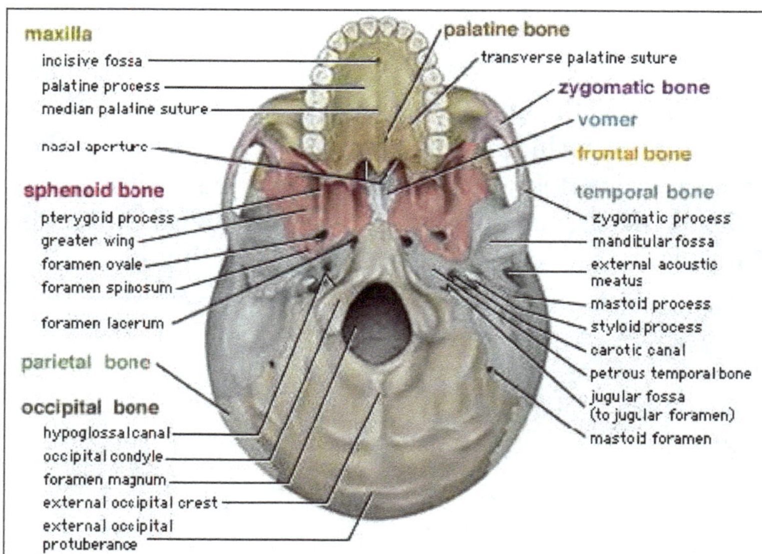

Inferior view of the human skull.

Anterior View of Skull

The anterior skull consists of the facial bones and provides the bony support for the eyes and structures of the face. This view of the skull is dominated by the openings of the orbits and the nasal cavity. Also seen are the upper and lower jaws, with their respective teeth.

The orbit is the bony socket that houses the eyeball and muscles that move the eyeball or open the upper eyelid. The upper margin of the anterior orbit is the supraorbital margin. Located near the midpoint of the supraorbital margin is a small opening called the supraorbital foramen. This provides for passage of a sensory nerve to the skin of the forehead. Below the orbit is the infraorbital foramen, which is the point of emergence for a sensory nerve that supplies the anterior face below the orbit.

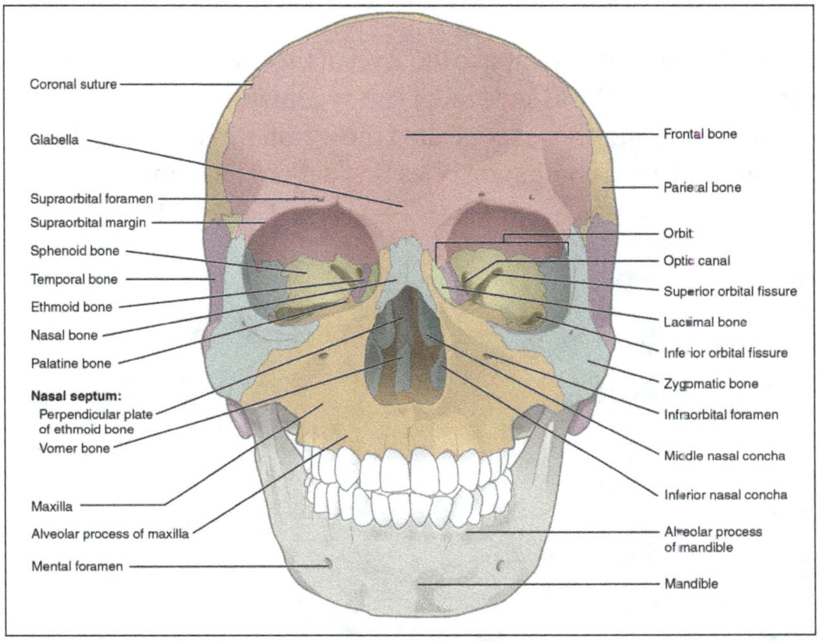

Anterior View of Skull. An anterior view of the skull shows the bones that form the forehead, orbits (eye sockets), nasal cavity, nasal septum, and upper and lower jaws.

Inside the nasal area of the skull, the nasal cavity is divided into halves by the nasal septum. The upper portion of the nasal septum is formed by the perpendicular plate of the ethmoid bone and the lower portion is the vomer bone. Each side of the nasal cavity is triangular in shape, with a broad inferior space that narrows superiorly. When looking into the nasal cavity from the front of the skull, two bony plates are seen projecting from each lateral wall. The larger of these is the inferior nasal concha, an independent bone of the skull. Located just above the inferior concha is the middle nasal concha, which is part of the ethmoid bone. A third bony plate, also part of the ethmoid bone, is the superior nasal concha. It is much smaller and out of sight, above the middle concha. The superior nasal concha is located just lateral to the perpendicular plate, in the upper nasal cavity.

Lateral View of Skull

A view of the lateral skull is dominated by the large, rounded brain case above and the upper and lower jaws with their teeth below. Separating these areas is the bridge of bone called the zygomatic arch. The zygomatic arch is the bony arch on the side of skull that spans from the area of the cheek to just above the ear canal. It is formed by the junction of two bony processes: a short anterior component, the temporal process of the zygomatic bone (the cheekbone) and a longer posterior portion, the zygomatic process of the temporal bone, extending forward from the temporal bone. Thus the temporal process (anteriorly) and the zygomatic process (posteriorly) join together, like the two ends of a drawbridge, to form the zygomatic arch. One of the major muscles that pulls the mandible upward during biting and chewing arises from the zygomatic arch.

On the lateral side of the brain case, above the level of the zygomatic arch, is a shallow space called the temporal fossa. Below the level of the zygomatic arch and deep to the vertical portion of the mandible is another space called the infratemporal fossa. Both the temporal fossa and infratemporal fossa contain muscles that act on the mandible during chewing.

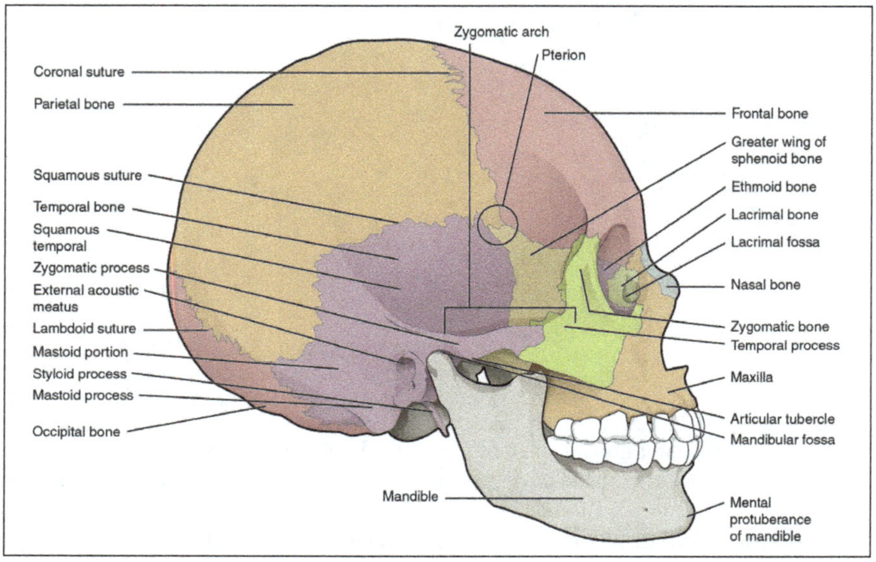

Lateral View of Skull.

The lateral skull shows the large rounded brain case, zygomatic arch, and the upper and lower jaws. The zygomatic arch is formed jointly by the zygomatic process of the temporal bone and the temporal process of the zygomatic bone. The shallow space above the zygomatic arch is the temporal fossa. The space inferior to the zygomatic arch and deep to the posterior mandible is the infratemporal fossa.

Bones of the Brain Case

The brain case contains and protects the brain. The interior space that is almost completely occupied by the brain is called the cranial cavity. This cavity is bounded superiorly by the rounded top of the skull, which is called the calvaria (skullcap), and the lateral and posterior sides of the skull. The bones that form the top and sides of the brain case are usually referred to as the "flat" bones of the skull.

The floor of the brain case is referred to as the base of the skull. This is a complex area that varies in depth and has numerous openings for the passage of cranial nerves, blood vessels, and the spinal cord. Inside the skull, the base is subdivided into three large spaces, called the anterior cranial fossa, middle cranial fossa, and posterior cranial fossa (fossa = "trench or ditch"). From anterior to posterior, the fossae increase in depth. The shape and depth of each fossa corresponds to the shape and size of the brain region that each houses.

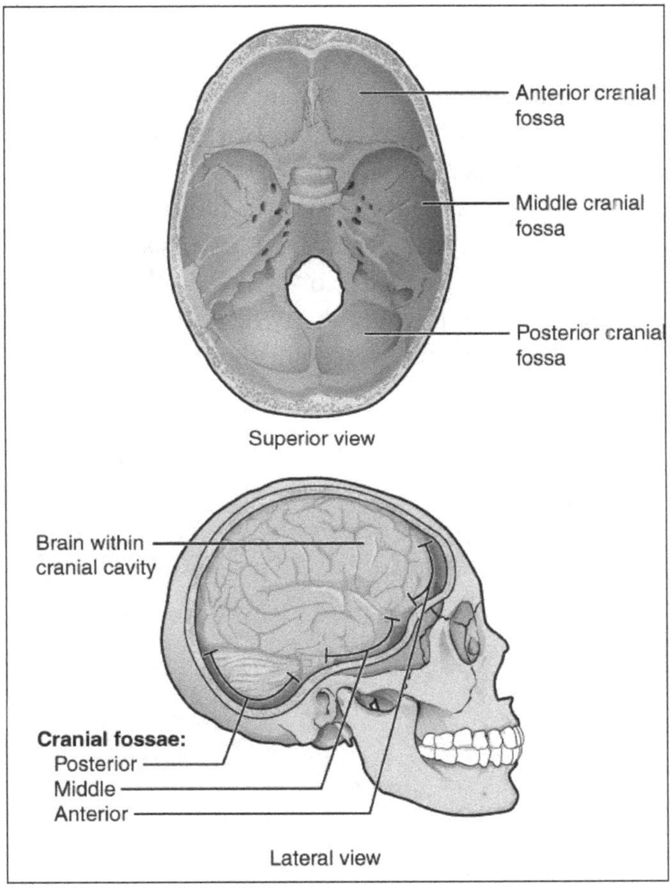

Cranial Fossae.

The bones of the brain case surround and protect the brain, which occupies the cranial cavity. The base of the brain case, which forms the floor of cranial cavity, is subdivided into the shallow anterior cranial fossa, the middle cranial fossa, and the deep posterior cranial fossa.

The brain case consists of eight bones. These include the paired parietal and temporal bones, plus the unpaired frontal, occipital, sphenoid, and ethmoid bones.

Parietal Bone

The parietal bone forms most of the upper lateral side of the skull. These are paired bones, with the right and left parietal bones joining together at the top of the skull. Each parietal bone is also bounded anteriorly by the frontal bone, inferiorly by the temporal bone, and posteriorly by the occipital bone.

Temporal Bone

The temporal bone forms the lower lateral side of the skull. Common wisdom has it that the temporal bone (temporal = "time") is so named because this area of the head (the temple) is where hair typically first turns gray, indicating the passage of time.

The temporal bone is subdivided into several regions. The flattened, upper portion is the squamous portion of the temporal bone. Below this area and projecting anteriorly is the zygomatic process of the temporal bone, which forms the posterior portion of the zygomatic arch. Posteriorly is the mastoid portion of the temporal bone. Projecting inferiorly from this region is a large prominence, the mastoid process, which serves as a muscle attachment site. The mastoid process can easily be felt on the side of the head just behind your earlobe. On the interior of the skull, the petrous portion of each temporal bone forms the prominent, diagonally oriented petrous ridge in the floor of the cranial cavity. Located inside each petrous ridge are small cavities that house the structures of the middle and inner ears.

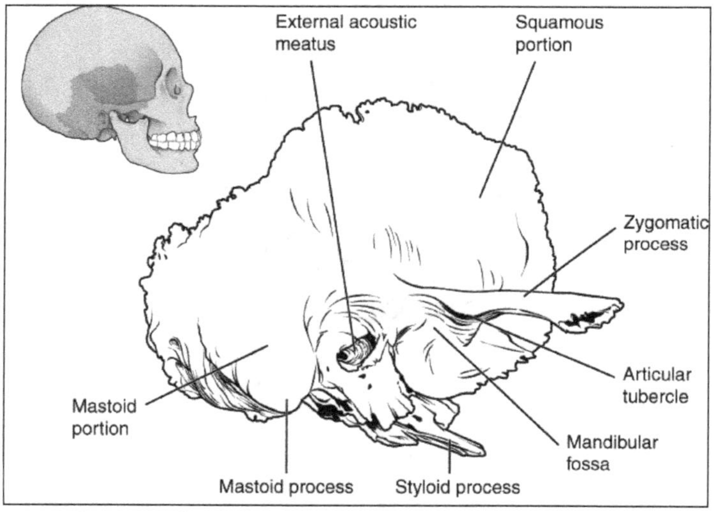

Temporal Bone: A lateral view of the isolated temporal bone shows the squamous, mastoid, and zygomatic portions of the temporal bone.

Important landmarks of the temporal bone, as shown in figure above, include the following:

- External acoustic meatus (ear canal): This is the large opening on the lateral side of the skull that is associated with the ear.
- Internal acoustic meatus: This opening is located inside the cranial cavity, on the medial side of the petrous ridge. It connects to the middle and inner ear cavities of the temporal bone.
- Mandibular fossa: This is the deep, oval-shaped depression located on the external base of the skull, just in front of the external acoustic meatus. The mandible

(lower jaw) joins with the skull at this site as part of the temporomandibular joint, which allows for movements of the mandible during opening and closing of the mouth.

- Articular tubercle: The smooth ridge located immediately anterior to the mandibular fossa. Both the articular tubercle and mandibular fossa contribute to the temporomandibular joint, the joint that provides for movements between the temporal bone of the skull and the mandible.
- Styloid process: Posterior to the mandibular fossa on the external base of the skull is an elongated, downward bony projection called the styloid process, so named because of its resemblance to a stylus (a pen or writing tool). This structure serves as an attachment site for several small muscles and for a ligament that supports the hyoid bone of the neck.

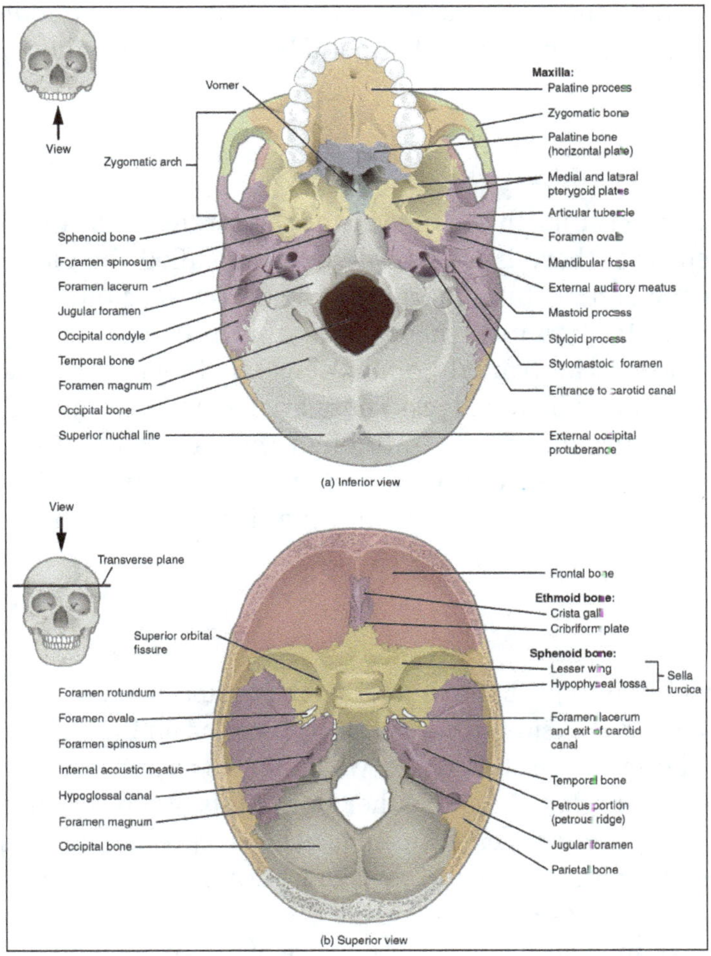

(a) Inferior view

(b) Superior view

- Stylomastoid foramen: This small opening is located between the styloid process and mastoid process. This is the point of exit for the cranial nerve that supplies the facial muscles.

- Carotid canal: The carotid canal is a zig-zag shaped tunnel that provides passage through the base of the skull for one of the major arteries that supplies the brain. Its entrance is located on the outside base of the skull, anteromedial to the styloid process. The canal then runs anteromedially within the bony base of the skull, and then turns upward to its exit in the floor of the middle cranial cavity, above the foramen lacerum.

External and internal views of base of skull:

- The hard palate is formed anteriorly by the palatine processes of the maxilla bones and posteriorly by the horizontal plate of the palatine bones.
- The complex floor of the cranial cavity is formed by the frontal, ethmoid, sphenoid, temporal, and occipital bones.

The lesser wing of the sphenoid bone separates the anterior and middle cranial fossae. The petrous ridge (petrous portion of temporal bone) separates the middle and posterior cranial fossae.

Frontal Bone

The frontal bone is the single bone that forms the forehead. At its anterior midline, between the eyebrows, there is a slight depression called the glabella. The frontal bone also forms the supraorbital margin of the orbit. Near the middle of this margin, is the supraorbital foramen, the opening that provides passage for a sensory nerve to the forehead. The frontal bone is thickened just above each supraorbital margin, forming rounded brow ridges. These are located just behind your eyebrows and vary in size among individuals, although they are generally larger in males. Inside the cranial cavity, the frontal bone extends posteriorly. This flattened region forms both the roof of the orbit below and the floor of the anterior cranial cavity above.

Occipital Bone

The occipital bone is the single bone that forms the posterior skull and posterior base of the cranial cavity. On its outside surface, at the posterior midline, is a small protrusion called the external occipital protuberance, which serves as an attachment site for a ligament of the posterior neck. Lateral to either side of this bump is a superior nuchal line (nuchal = "nape" or "posterior neck"). The nuchal lines represent the most superior point at which muscles of the neck attach to the skull, with only the scalp covering the skull above these lines. On the base of the skull, the occipital bone contains the large opening of the foramen magnum, which allows for passage of the spinal cord as it exits the skull. On either side of the foramen magnum is an oval-shaped occipital condyle. These condyles form joints with the first cervical vertebra and thus support the skull on top of the vertebral column.

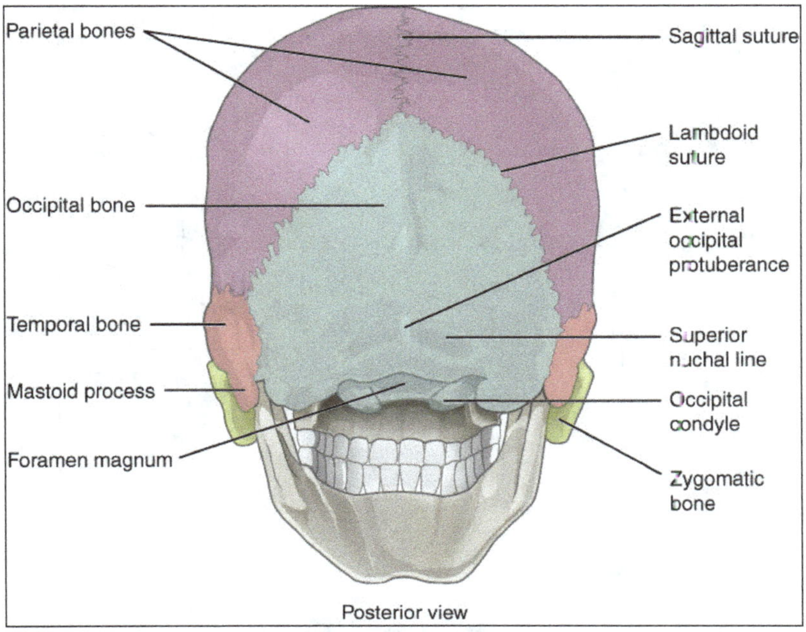

Posterior view of Skull: This view of the posterior skull shows attachment sites for muscles and joints that support the skull.

Sphenoid Bone

The sphenoid bone is a single, complex bone of the central skull. It serves as a "keystone" bone, because it joins with almost every other bone of the skull. The sphenoid forms much of the base of the central skull and also extends laterally to contribute to the sides of the skull. Inside the cranial cavity, the right and left lesser wings of the sphenoid bone, which resemble the wings of a flying bird, form the lip of a prominent ridge that marks the boundary between the anterior and middle cranial fossae. The sella turcica ("Turkish saddle") is located at the midline of the middle cranial fossa. This bony region of the sphenoid bone is named for its resemblance to the horse saddles used by the Ottoman Turks, with a high back and a tall front. The rounded depression in the floor of the sella turcica is the hypophyseal (pituitary) fossa, which houses the pea-sized pituitary (hypophyseal) gland. The greater wings of the sphenoid bone extend laterally to either side away from the sella turcica, where they form the anterior floor of the middle cranial fossa. The greater wing is best seen on the outside of the lateral skull, where it forms a rectangular area immediately anterior to the squamous portion of the temporal bone.

On the inferior aspect of the skull, each half of the sphenoid bone forms two thin, vertically oriented bony plates. These are the medial pterygoid plate and lateral pterygoid plate (pterygoid = "wing-shaped"). The right and left medial pterygoid plates form the posterior, lateral walls of the nasal cavity. The somewhat larger lateral pterygoid plates serve as attachment sites for chewing muscles that fill the infratemporal space and act on the mandible.

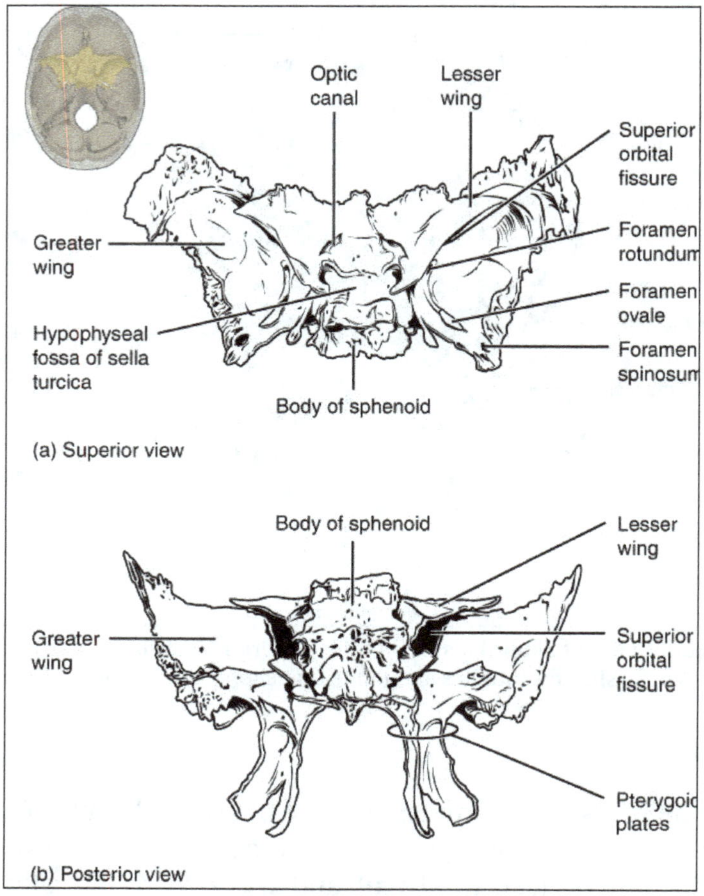

Sphenoid Bone.

Shown in isolation in (a) superior and (b) posterior views, the sphenoid bone is a single midline bone that forms the anterior walls and floor of the middle cranial fossa. It has a pair of lesser wings and a pair of greater wings. The sella turcica surrounds the hypophyseal fossa. Projecting downward are the medial and lateral pterygoid plates. The sphenoid has multiple openings for the passage of nerves and blood vessels, including the optic canal, superior orbital fissure, foramen rotundum, foramen ovale, and foramen spinosum.

Ethmoid Bone

The ethmoid bone is a single, midline bone that forms the roof and lateral walls of the upper nasal cavity, the upper portion of the nasal septum, and contributes to the medial wall of the orbit. On the interior of the skull, the ethmoid also forms a portion of the floor of the anterior cranial cavity.

Within the nasal cavity, the perpendicular plate of the ethmoid bone forms the upper portion of the nasal septum. The ethmoid bone also forms the lateral walls of the upper nasal cavity. Extending from each lateral wall are the superior nasal concha and middle nasal concha, which are thin, curved projections that extend into the nasal cavity.

In the cranial cavity, the ethmoid bone forms a small area at the midline in the floor of the anterior cranial fossa. This region also forms the narrow roof of the underlying nasal cavity. This portion of the ethmoid bone consists of two parts, the crista galli and cribriform plates. The crista galli ("rooster's comb or crest") is a small upward bony projection located at the midline. It functions as an anterior attachment point for one of the covering layers of the brain. To either side of the crista galli is the cribriform plate (cribrum = "sieve"), a small, flattened area with numerous small openings termed olfactory foramina. Small nerve branches from the olfactory areas of the nasal cavity pass through these openings to enter the brain.

The lateral portions of the ethmoid bone are located between the orbit and upper nasal cavity, and thus form the lateral nasal cavity wall and a portion of the medial orbit wall. Located inside this portion of the ethmoid bone are several small, air-filled spaces that are part of the paranasal sinus system of the skull.

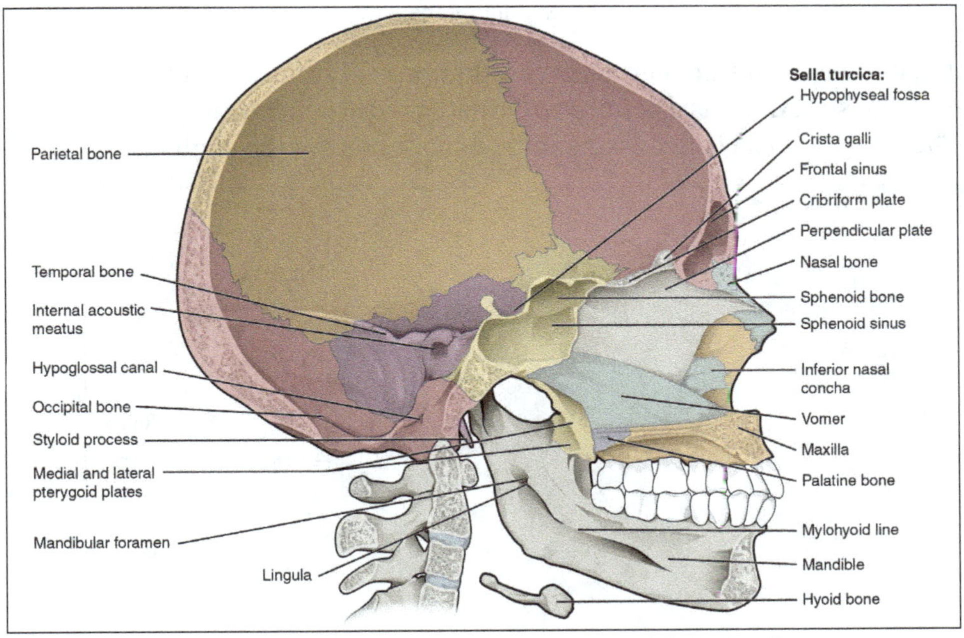

Sagittal Section of Skull. This midline view of the sagittally sectioned skull shows the nasal septum.

The unpaired ethmoid bone is located at the midline within the central skull. It has an upward projection, the crista galli, and a downward projection, the perpendicular plate, which forms the upper nasal septum. The cribriform plates form both the roof of the nasal cavity and a portion of the anterior cranial fossa floor. The lateral sides of the ethmoid bone form the lateral walls of the upper nasal cavity, part of the medial orbit wall, and give rise to the superior and middle nasal conchae. The ethmoid bone also contains the ethmoid air cells.

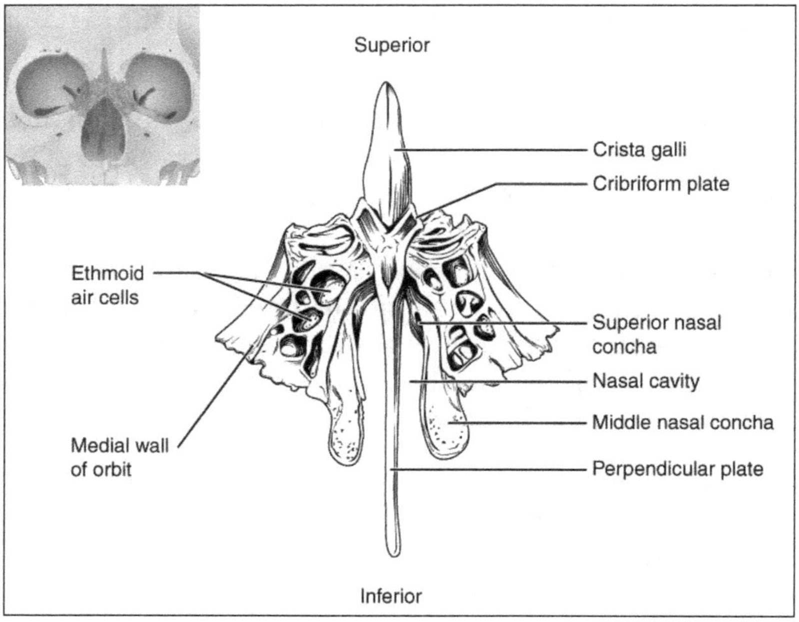

Ethmoid Bone.

The three nasal conchae are curved bones that project from the lateral walls of the nasal cavity. The superior nasal concha and middle nasal concha are parts of the ethmoid bone. The inferior nasal concha is an independent bone of the skull.

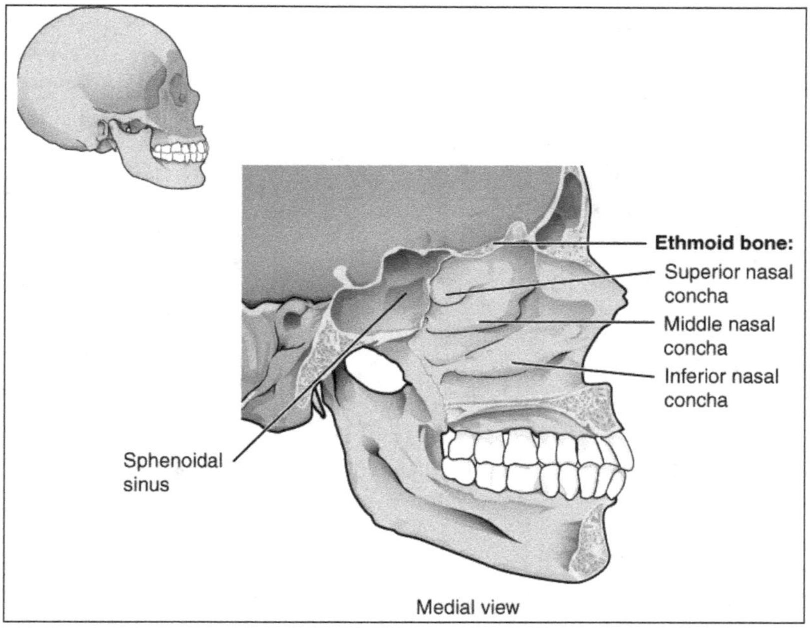

Lateral Wall of Nasal Cavity.

Sutures of the Skull

A suture is an immobile joint between adjacent bones of the skull. The narrow gap between the bones is filled with dense, fibrous connective tissue that unites the bones. The long sutures located between the bones of the brain case are not straight, but instead follow irregular, tightly twisting paths. These twisting lines serve to tightly interlock the adjacent bones, thus adding strength to the skull for brain protection.

The two suture lines seen on the top of the skull are the coronal and sagittal sutures. The coronal suture runs from side to side across the skull, within the coronal plane of section. It joins the frontal bone to the right and left parietal bones. The sagittal suture extends posteriorly from the coronal suture, running along the midline at the top of the skull in the sagittal plane of section. It unites the right and left parietal bones. On the posterior skull, the sagittal suture terminates by joining the lambdoid suture. The lambdoid suture extends downward and laterally to either side away from its junction with the sagittal suture. The lambdoid suture joins the occipital bone to the right and left parietal and temporal bones. This suture is named for its upside-down "V" shape, which resembles the capital letter version of the Greek letter lambda (Λ). The squamous suture is located on the lateral skull. It unites the squamous portion of the temporal bone with the parietal bone. At the intersection of four bones is the pterion, a small, capital-H-shaped suture line region that unites the frontal bone, parietal bone, squamous portion of the temporal bone, and greater wing of the sphenoid bone. It is the weakest part of the skull. The pterion is located approximately two finger widths above the zygomatic arch and a thumb's width posterior to the upward portion of the zygomatic bone.

Skeletal System

Head and traumatic brain injuries are major causes of immediate death and disability, with bleeding and infections as possible additional complications. According to the Centers for Disease Control and Prevention, approximately 30 percent of all injury-related deaths in the United States are caused by head injuries. The majority of head injuries involve falls. They are most common among young children (ages 0–4 years), adolescents (15–19 years), and the elderly (over 65 years). Additional causes vary, but prominent among these are automobile and motorcycle accidents.

Strong blows to the brain-case portion of the skull can produce fractures. These may result in bleeding inside the skull with subsequent injury to the brain. The most common is a linear skull fracture, in which fracture lines radiate from the point of impact. Other fracture types include a comminuted fracture, in which the bone is broken into several pieces at the point of impact, or a depressed fracture, in which the fractured bone is pushed inward. In a contrecoup (counterblow) fracture, the bone at the point of impact is not broken, but instead a fracture occurs on the opposite side of the skull. Fractures of the occipital bone at the base of the skull can occur in this manner, producing a basilar fracture that can damage the artery that passes through the carotid canal.

A blow to the lateral side of the head may fracture the bones of the pterion. The pterion is an important clinical landmark because located immediately deep to it on the inside of the skull is a major branch of an artery that supplies the skull and covering layers of the brain. A strong blow to this region can fracture the bones around the pterion. If the underlying artery is damaged, bleeding can cause the formation of a hematoma (collection of blood) between the brain and interior of the skull. As blood accumulates, it will put pressure on the brain. Symptoms associated with a hematoma may not be apparent immediately following the injury, but if untreated, blood accumulation will exert increasing pressure on the brain and can result in death within a few hours.

Facial Bones of the Skull

The facial bones of the skull form the upper and lower jaws, the nose, nasal cavity and nasal septum, and the orbit. The facial bones include 14 bones, with six paired bones and two unpaired bones. The paired bones are the maxilla, palatine, zygomatic, nasal, lacrimal, and inferior nasal conchae bones. The unpaired bones are the vomer and mandible bones. Although classified with the brain-case bones, the ethmoid bone also contributes to the nasal septum and the walls of the nasal cavity and orbit.

Maxillary Bone

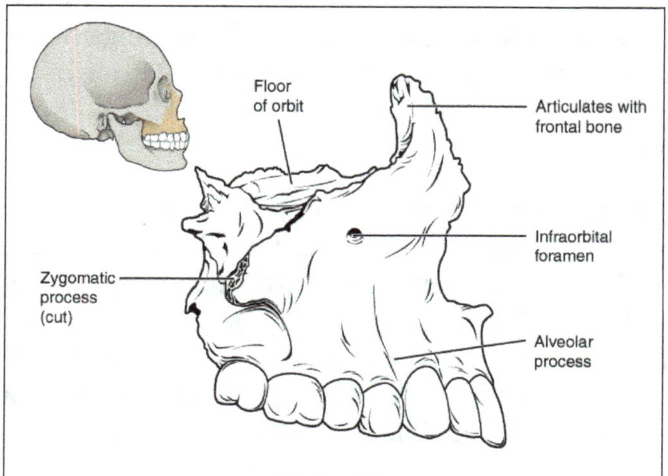

Maxillary Bone: The maxillary bone forms the upper jaw and supports the upper teeth. Each maxilla also forms the lateral floor of each orbit and the majority of the hard palate.

The maxillary bone, often referred to simply as the maxilla (plural = maxillae), is one of a pair that together form the upper jaw, much of the hard palate, the medial floor of the orbit, and the lateral base of the nose. The curved, inferior margin of the maxillary bone that forms the upper jaw and contains the upper teeth is the alveolar process of the maxilla. Each tooth is anchored into a deep socket called an alveolus. On the anterior maxilla, just below the orbit, is the infraorbital foramen. This is the point of exit for a sensory nerve that supplies the nose, upper lip, and anterior cheek. On the inferior skull, the palatine process from each maxillary bone can be seen joining together at the midline to form the anterior three-quarters of the hard palate. The hard palate is the bony plate that forms the roof of the mouth and floor of the nasal cavity, separating the oral and nasal cavities.

Palatine Bone

The palatine bone is one of a pair of irregularly shaped bones that contribute small areas to the lateral walls of the nasal cavity and the medial wall of each orbit. The largest region of each of the palatine bone is the horizontal plate. The plates from the right and left palatine bones join together at the midline to form the posterior quarter of the hard palate. Thus, the palatine bones are best seen in an inferior view of the skull and hard palate.

Homeostatic Imbalances

Cleft Lip and Cleft Palate

During embryonic development, the right and left maxilla bones come together at the midline to form the upper jaw. At the same time, the muscle and skin overlying these bones join together to form the upper lip. Inside the mouth, the palatine processes of the maxilla bones, along with the horizontal plates of the right and left palatine bones, join together to form the hard palate. If an error occurs in these developmental processes, a birth defect of cleft lip or cleft palate may result.

Cleft lip is a common development defect that affects approximately 1:1000 births, most of which are male. This defect involves a partial or complete failure of the right and left portions of the upper lip to fuse together, leaving a cleft (gap).

A more severe developmental defect is cleft palate, which affects the hard palate. The hard palate is the bony structure that separates the nasal cavity from the oral cavity. It is formed during embryonic development by the midline fusion of the horizontal plates from the right and left palatine bones and the palatine processes of the maxilla bones. Cleft palate affects approximately 1:2500 births and is more common in females. It results from a failure of the two halves of the hard palate to completely come together and fuse at the midline, thus leaving a gap between them. This gap allows for communication between the nasal and oral cavities. In severe cases, the bony gap continues into the anterior upper jaw where the alveolar processes of the maxilla bones also do not properly join together above the front teeth. If this occurs, a cleft lip will also be seen. Because of the communication between the oral and nasal cavities, a cleft palate makes it very difficult for an infant to generate the suckling needed for nursing, thus leaving the infant at risk for malnutrition. Surgical repair is required to correct cleft palate defects.

Zygomatic Bone

The zygomatic bone is also known as the cheekbone. Each of the paired zygomatic bones forms much of the lateral wall of the orbit and the lateral-inferior margins of the anterior orbital opening. The short temporal process of the zygomatic bone projects posteriorly, where it forms the anterior portion of the zygomatic arch.

Nasal Bone

The nasal bone is one of two small bones that articulate (join) with each other to form the bony base (bridge) of the nose. They also support the cartilages that form the lateral walls of the nose. These are the bones that are damaged when the nose is broken.

Lacrimal Bone

Each lacrimal bone is a small, rectangular bone that forms the anterior, medial wall of the orbit. The anterior portion of the lacrimal bone forms a shallow depression called the lacrimal fossa, and extending inferiorly from this is the nasolacrimal canal. The lacrimal fluid (tears of the eye), which serves to maintain the moist surface of the eye, drains at the medial corner of the eye into the nasolacrimal canal. This duct then extends downward to open into the nasal cavity, behind the inferior nasal concha. In the nasal cavity, the lacrimal fluid normally drains posteriorly, but with an increased flow of tears due to crying or eye irritation, some fluid will also drain anteriorly, thus causing a runny nose.

Inferior Nasal Conchae

The right and left inferior nasal conchae form a curved bony plate that projects into the nasal cavity space from the lower lateral wall. The inferior concha is the largest of the nasal conchae and can easily be seen when looking into the anterior opening of the nasal cavity.

Vomer Bone

The unpaired vomer bone, often referred to simply as the vomer, is triangular-shaped and forms the posterior-inferior part of the nasal septum. The vomer is best seen when looking from behind into the posterior openings of the nasal cavity. In this view, the vomer is seen to form the entire height of the nasal septum. A much smaller portion of the vomer can also be seen when looking into the anterior opening of the nasal cavity.

Mandible

The mandible forms the lower jaw and is the only moveable bone of the skull. At the time of birth, the mandible consists of paired right and left bones, but these fuse together during the first year to form the single U-shaped mandible of the adult skull. Each side of the mandible consists of a horizontal body and posteriorly, a vertically oriented ramus of the mandible (ramus = “branch”). The outside margin of the mandible, where the body and ramus come together is called the angle of the mandible.

The ramus on each side of the mandible has two upward-going bony projections. The more anterior projection is the flattened coronoid process of the mandible, which provides attachment for one of the biting muscles. The posterior projection is the condylar process of the mandible, which is topped by the oval-shaped condyle. The condyle of the mandible articulates (joins) with the mandibular fossa and articular tubercle of the temporal bone. Together these articulations form the temporomandibular joint, which allows for opening and closing of the mouth. The broad U-shaped curve located between the coronoid and condylar processes is the mandibular notch.

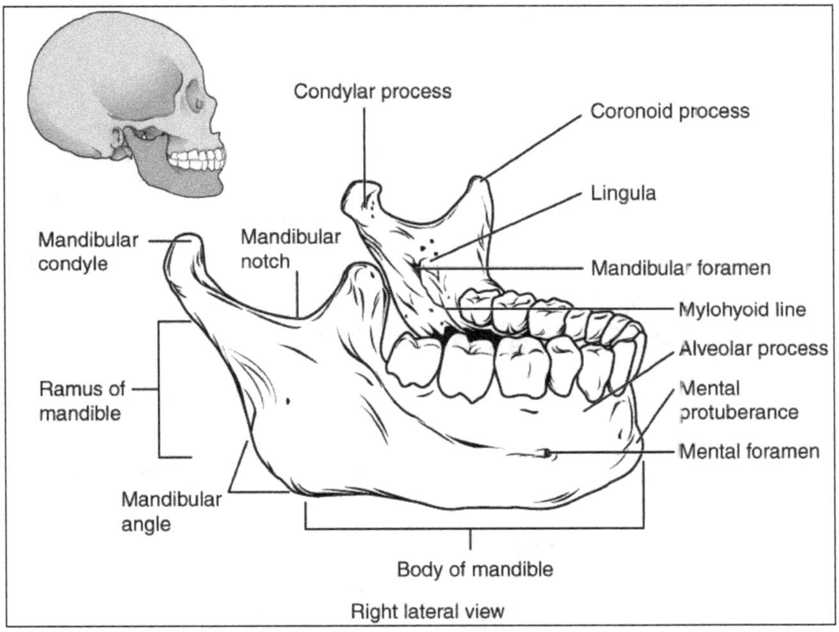

Isolated Mandible. The mandible is the only moveable bone of the skull.

Important landmarks for the mandible include the following:

- Alveolar process of the mandible: This is the upper border of the mandibular body and serves to anchor the lower teeth.
- Mental protuberance: The forward projection from the inferior margin of the anterior mandible that forms the chin (mental = "chin").
- Mental foramen: The opening located on each side of the anterior-lateral mandible, which is the exit site for a sensory nerve that supplies the chin.
- Mylohyoid line: This bony ridge extends along the inner aspect of the mandibular body. The muscle that forms the floor of the oral cavity attaches to the mylohyoid lines on both sides of the mandible.
- Mandibular foramen: This opening is located on the medial side of the ramus of the mandible. The opening leads into a tunnel that runs down the length of

the mandibular body. The sensory nerve and blood vessels that supply the lower teeth enter the mandibular foramen and then follow this tunnel. Thus, to numb the lower teeth prior to dental work, the dentist must inject anesthesia into the lateral wall of the oral cavity at a point prior to where this sensory nerve enters the mandibular foramen.

- Lingula: This small flap of bone is named for its shape (lingula = "little tongue"). It is located immediately next to the mandibular foramen, on the medial side of the ramus. A ligament that anchors the mandible during opening and closing of the mouth extends down from the base of the skull and attaches to the lingula.

The Orbit

The orbit is the bony socket that houses the eyeball and contains the muscles that move the eyeball or open the upper eyelid. Each orbit is cone-shaped, with a narrow posterior region that widens toward the large anterior opening. To help protect the eye, the bony margins of the anterior opening are thickened and somewhat constricted. The medial walls of the two orbits are parallel to each other but each lateral wall diverges away from the midline at a 45° angle. This divergence provides greater lateral peripheral vision.

The walls of each orbit include contributions from seven skull bones. The frontal bone forms the roof and the zygomatic bone forms the lateral wall and lateral floor. The medial floor is primarily formed by the maxilla, with a small contribution from the palatine bone. The ethmoid bone and lacrimal bone make up much of the medial wall and the sphenoid bone forms the posterior orbit.

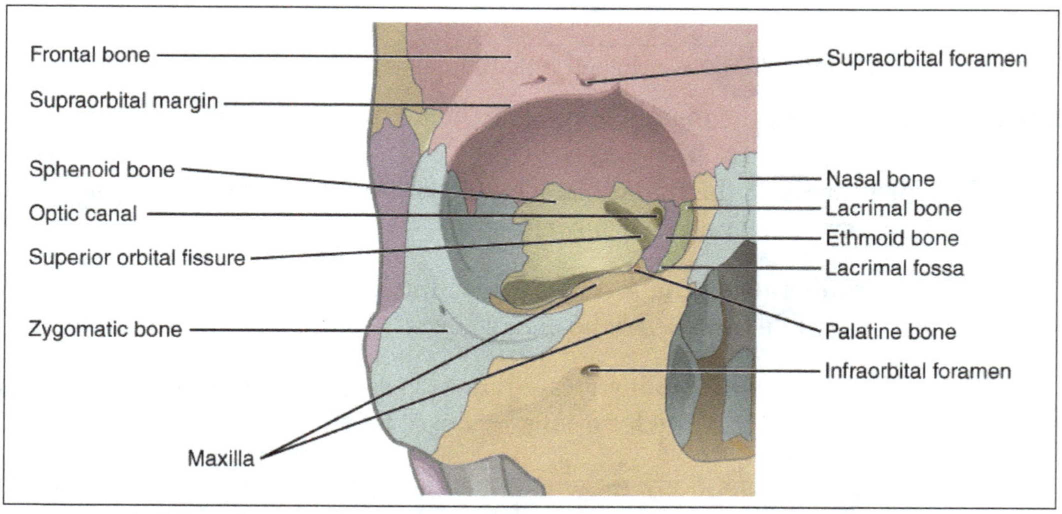

Bones of the Orbit: Seven skull bones contribute to the walls of the orbit. Opening into the posterior orbit from the cranial cavity are the optic canal and superior orbital fissure.

At the posterior apex of the orbit is the opening of the optic canal, which allows for passage of the optic nerve from the retina to the brain. Lateral to this is the elongated and irregularly shaped superior orbital fissure, which provides passage for the artery that supplies the eyeball, sensory nerves, and the nerves that supply the muscles involved in eye movements.

The Nasal Septum and Nasal Conchae

The nasal septum consists of both bone and cartilage components. The upper portion of the septum is formed by the perpendicular plate of the ethmoid bone. The lower and posterior parts of the septum are formed by the triangular-shaped vomer bone. In an anterior view of the skull, the perpendicular plate of the ethmoid bone is easily seen inside the nasal opening as the upper nasal septum, but only a small portion of the vomer is seen as the inferior septum. A better view of the vomer bone is seen when looking into the posterior nasal cavity with an inferior view of the skull, where the vomer forms the full height of the nasal septum. The anterior nasal septum is formed by the septal cartilage, a flexible plate that fills in the gap between the perpendicular plate of the ethmoid and vomer bones. This cartilage also extends outward into the nose where it separates the right and left nostrils. The septal cartilage is not found in the dry skull.

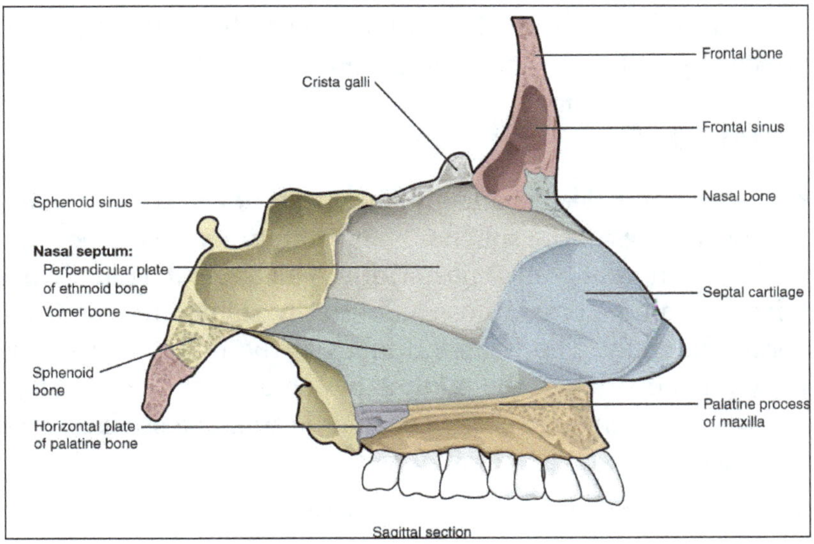

Nasal Septum: The nasal septum is formed by the perpendicular plate of the ethmoid bone and the vomer bone. The septal cartilage fills the gap between these bones and extends into the nose.

Attached to the lateral wall on each side of the nasal cavity are the superior, middle, and inferior nasal conchae (singular = concha), which are named for their positions. These are bony plates that curve downward as they project into the space of the nasal cavity. They serve to swirl the incoming air, which helps to warm and moisturize it before the air moves into the delicate air sacs of the lungs. This also allows mucus, secreted by the tissue lining the nasal cavity, to trap incoming dust, pollen, bacteria, and viruses. The largest of the conchae is the inferior nasal concha, which is an independent bone of the skull. The middle concha and the superior conchae, which is the smallest, are both formed by the ethmoid bone. When looking into the anterior nasal opening of the skull, only the inferior and middle conchae can be seen. The small superior nasal concha is well hidden above and behind the middle concha.

Cranial Fossae

Inside the skull, the floor of the cranial cavity is subdivided into three cranial fossae (spaces), which increase in depth from anterior to posterior. Since the brain occupies these areas, the shape of each conforms to the shape of the brain regions that it contains. Each cranial fossa has anterior and posterior boundaries and is divided at the midline into right and left areas by a significant bony structure or opening.

Anterior Cranial Fossa

The anterior cranial fossa is the most anterior and the shallowest of the three cranial fossae. It overlies the orbits and contains the frontal lobes of the brain. Anteriorly, the anterior fossa is bounded by the frontal bone, which also forms the majority of the floor for this space. The lesser wings of the sphenoid bone form the prominent ledge that marks the boundary between the anterior and middle cranial fossae. Located in the floor of the anterior cranial fossa at the midline is a portion of the ethmoid bone, consisting of the upward projecting crista galli and to either side of this, the cribriform plates.

Middle Cranial Fossa

The middle cranial fossa is deeper and situated posterior to the anterior fossa. It extends from the lesser wings of the sphenoid bone anteriorly, to the petrous ridges (petrous portion of the temporal bones) posteriorly. The large, diagonally positioned petrous ridges give the middle cranial fossa a butterfly shape, making it narrow at the midline and broad laterally. The temporal lobes of the brain occupy this fossa. The middle cranial fossa is divided at the midline by the upward bony prominence of the sella turcica, a part of the sphenoid bone. The middle cranial fossa has several openings for the passage of blood vessels and cranial nerves.

Openings in the middle cranial fossa are as follows:

- Optic canal: This opening is located at the anterior lateral corner of the sella turcica. It provides for passage of the optic nerve into the orbit.
- Superior orbital fissure: This large, irregular opening into the posterior orbit is located on the anterior wall of the middle cranial fossa, lateral to the optic canal and under the projecting margin of the lesser wing of the sphenoid bone. Nerves to the eyeball and associated muscles, and sensory nerves to the forehead pass through this opening.
- Foramen rotundum: This rounded opening (rotundum = "round") is located in the floor of the middle cranial fossa, just inferior to the superior orbital fissure.

It is the exit point for a major sensory nerve that supplies the cheek, nose, and upper teeth.

- Foramen ovale of the middle cranial fossa: This large, oval-shaped opening in the floor of the middle cranial fossa provides passage for a major sensory nerve to the lateral head, cheek, chin, and lower teeth.
- Foramen spinosum: This small opening, located posterior-lateral to the foramen ovale, is the entry point for an important artery that supplies the covering layers surrounding the brain. The branching pattern of this artery forms readily visible grooves on the internal surface of the skull and these grooves can be traced back to their origin at the foramen spinosum.
- Carotid canal: This is the zig-zag passageway through which a major artery to the brain enters the skull. The entrance to the carotid canal is located on the inferior aspect of the skull, anteromedial to the styloid process. From here, the canal runs anteromedially within the bony base of the skull. Just above the foramen lacerum, the carotid canal opens into the middle cranial cavity, near the posterior-lateral base of the sella turcica.
- Foramen lacerum: This irregular opening is located in the base of the skull, immediately inferior to the exit of the carotid canal. This opening is an artifact of the dry skull, because in life it is completely filled with cartilage. All the openings of the skull that provide for passage of nerves or blood vessels have smooth margins; the word lacerum ("ragged" or "torn") tells us that this opening has ragged edges and thus nothing passes through it.

Posterior Cranial Fossa

The posterior cranial fossa is the most posterior and deepest portion of the cranial cavity. It contains the cerebellum of the brain. The posterior fossa is bounded anteriorly by the petrous ridges, while the occipital bone forms the floor and posterior wall. It is divided at the midline by the large foramen magnum ("great aperture"), the opening that provides for passage of the spinal cord.

Located on the medial wall of the petrous ridge in the posterior cranial fossa is the internal acoustic meatus. This opening provides for passage of the nerve from the hearing and equilibrium organs of the inner ear, and the nerve that supplies the muscles of the face. Located at the anterior-lateral margin of the foramen magnum is the hypoglossal canal. These emerge on the inferior aspect of the skull at the base of the occipital condyle and provide passage for an important nerve to the tongue.

Immediately inferior to the internal acoustic meatus is the large, irregularly shaped jugular foramen. Several cranial nerves from the brain exit the skull via this opening. It is also the exit point through the base of the skull for all the venous return blood leaving the brain. The venous structures that carry blood inside the skull form large, curved grooves on the inner walls of the posterior cranial fossa, which terminate at each jugular foramen.

Paranasal Sinuses

The paranasal sinuses are hollow, air-filled spaces located within certain bones of the skull. All of the sinuses communicate with the nasal cavity (paranasal = "next to nasal cavity") and are lined with nasal mucosa. They serve to reduce bone mass and thus lighten the skull, and they also add resonance to the voice. This second feature is most obvious when you have a cold or sinus congestion. These produce swelling of the mucosa and excess mucus production, which can obstruct the narrow passageways between the sinuses and the nasal cavity, causing your voice to sound different to yourself and others. This blockage can also allow the sinuses to fill with fluid, with the resulting pressure producing pain and discomfort.

The paranasal sinuses are named for the skull bone that each occupies. The frontal sinus is located just above the eyebrows, within the frontal bone. This irregular space may be divided at the midline into bilateral spaces, or these may be fused into a single sinus space. The frontal sinus is the most anterior of the paranasal sinuses. The largest sinus is the maxillary sinus. These are paired and located within the right and left maxillary bones, where they occupy the area just below the orbits. The maxillary sinuses are most commonly involved during sinus infections. Because their connection to the nasal cavity is located high on their medial wall, they are difficult to drain. The sphenoid sinus is a single, midline sinus. It is located within the body of the sphenoid bone, just anterior and inferior to the sella turcica, thus making it the most posterior of the paranasal sinuses. The lateral aspects of the ethmoid bone contain multiple small spaces separated by very thin bony walls. Each of these spaces is called an ethmoid air cell. These are located on both sides of the ethmoid bone, between the upper nasal cavity and medial orbit, just behind the superior nasal conchae.

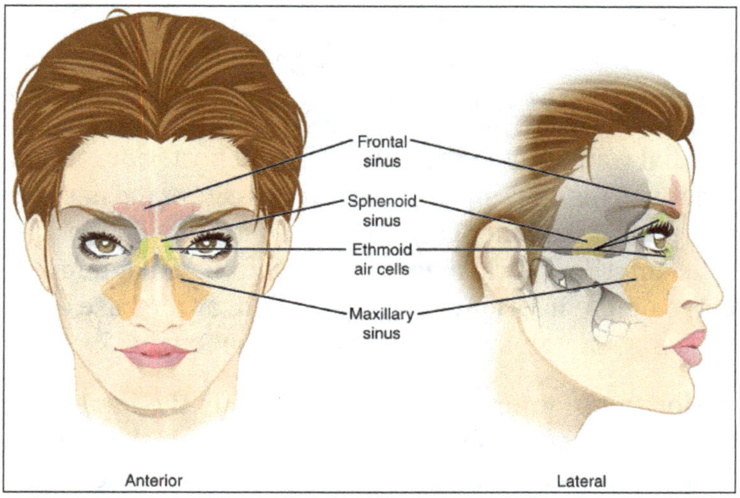

Paranasal Sinuses.

The paranasal sinuses are hollow, air-filled spaces named for the skull bone that each occupies. The most anterior is the frontal sinus, located in the frontal bone above the eyebrows. The largest are the maxillary sinuses, located in the right and left maxillary bones below the orbits. The most posterior is the sphenoid sinus, located in the body of the sphenoid bone, under the sella turcica. The ethmoid air cells are multiple small spaces located in the right and left sides of the ethmoid bone, between the medial wall of the orbit and lateral wall of the upper nasal cavity.

Hyoid Bone

The hyoid bone is an independent bone that does not contact any other bone and thus is not part of the skull. It is a small U-shaped bone located in the upper neck near the level of the inferior mandible, with the tips of the "U" pointing posteriorly. The hyoid serves as the base for the tongue above, and is attached to the larynx below and the pharynx posteriorly. The hyoid is held in position by a series of small muscles that attach to it either from above or below. These muscles act to move the hyoid up/down or forward/back. Movements of the hyoid are coordinated with movements of the tongue, larynx, and pharynx during swallowing and speaking.

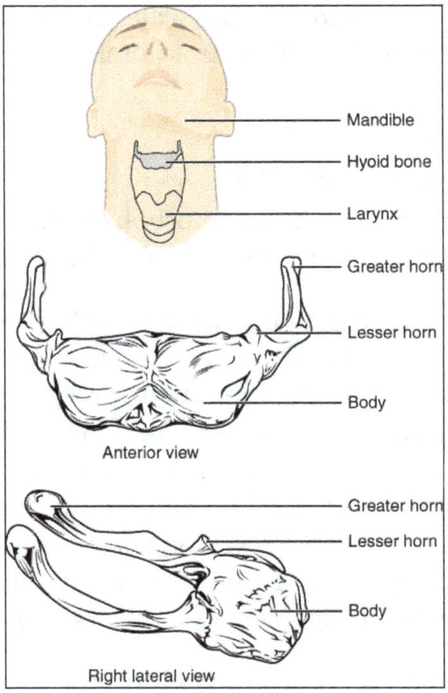

The hyoid bone is located in the upper neck and does not join with any other bone. It provides attachments for muscles that act on the tongue, larynx, and pharynx.

Cervical Nerves

Cervical nerves are spinal nerves that arise from the cervical region of the spinal cord. These nerves conduct motor and sensory information via efferent and afferent fibers, respectively, to and from the central nervous system. While classified as peripheral nerves, the motor cell body resides in the anterior horn of the spinal cord. There are eight pairs of cervical nerves, denoted C1 to C8, that emerge from the spinal cord superior to their corresponding vertebrae, except for C8 that exits inferiorly to the C7 vertebra. These nerves intertwine through plexuses that give rise to peripheral nerves that maintain a significant motor function in the head, neck, upper limbs and diaphragm, as well as sensation in the head, neck, shoulders and upper limbs in a dermatomal pattern.

Structure and Function

The cervical nerves arise from the spinal cord in the form of rootlets, or fila radicularia, smaller neuron bundles that coalesce to form roots. For each spinal nerve, an anterior and posterior root join to form the completed nerve. Shortly after branching out of the spinal cord, the cervical nerves form the cervical and brachial plexuses. It is important to note that the inputs and outputs may vary by individual and therefore, different sources may offer different descriptions.

The cervical plexus forms from the ventral rami of C1 to C4. It is known to anastomose with the facial nerve, hypoglossal nerve, spinal accessory nerve, vagus nerve, and the sympathetic trunk. It lies anteromedial to the scalenes, but is deep to the sternocleidomastoid, and gives rise to the motor and sensory branches:

Motor

- The C1 spinal nerve gives rise to the nerve to the geniohyoid and the nerve to the thyrohyoid that both function in the oropharynx. Anatomically, these nerves travel closely to the hypoglossal nerve (CN XII).
- The ansa cervicalis ("handle of the neck" in Latin), is a loop of nerves that lies superficial to the internal jugular vein, composed of the C1 to C3 nerves. More specifically, one end of the loop, the superior root, is derived from C1 (and possibly C2, depending on the literature), while the other, the inferior root, comes from C2 and C3. This structure gives rise to the nerves to the sternohyoid, sternothyroid, and superior and inferior bellies of the omohyoid muscles. All of the muscles supplied by branches of the ansa cervicalis, along with the thyrohyoid, are infrahyoid muscles, and therefore serve to depress the hyoid bone. This function is necessary for proper speech and swallowing.
- The C3, C4, and C5 roots contribute to form the phrenic nerves. These are responsible for motor and sensory supply to the diaphragm, as well as for sympathetic output. The phrenic nerves run anteriorly to the subclavian arteries, but posterior to the subclavian veins.
- Outside of the cervical plexus, the several muscles are innervated directly by upper cervical nerve roots:
    - C1 to C2: Rectus capitis anterior/lateralis.

- C1 to C3/4/5: Longus capitis.
- C2/3 to C4: Trapezius (sensory).

Sensory

The cutaneous branches of the cervical plexus transmit sensory information from the skin of the neck, the superior portions of the thorax, and the scalp. These nerves emerge from a common point at the posterior aspect of the sternocleidomastoid, known as Erb's point. There are several branches:

- The lesser occipital nerve arises from C2 and supplies the skin of the neck and the scalp posterior and superior to the auricle.
- The greater auricular nerve, of C2 and C3, ascends anterior to the auricle and supplies the skin over the parotid gland, the mastoid process and the skin from the parotid gland to the mastoid process.
- The transverse cervical nerve, of C2 and C3, projects posteriorly and wraps around anteriorly to supply the skin of the anterior cervical region.
- The supraclavicular nerve, of C3 and C4, descend from the posterior sternocleidomastoid and provides sensory information from the skin over the clavicle and shoulder.

Sympathetics

The ventral rami that compose the cervical plexus each receive gray rami communicantes from the superior cervical ganglion of the sympathetic trunk.

The brachial plexus forms from the ventral rami of C5 to C8, as well as T1. The plexus projects laterally, anterior to the first rib, but posterior to the clavicle, into the axillary region. The plexus is divided into several sections through its several anastomoses from the five nerve roots into three trunks, six divisions, three cords, and finally, five branches. The final outputs are best described by the location from which they emerge from the plexus, rather than simply based on their motor/sensory function as several nerves carry both types of information:

- Roots:
    - The C4 and C5 roots give rise to the dorsal scapular nerve that supplies the rhomboids and levator scapulae.
    - C5, as mentioned earlier, along with C3 and C4, contributes to the phrenic nerve that innervates the diaphragm.

    - Roots C5, C6, and C7 produce the long thoracic nerve, responsible for controlling the serratus anterior.
- Trunks:
    - From the upper trunk C5 and C6 give rise to the nerve to the subclavius, and the suprascapular nerve, that supply the subclavius muscle, and the supra- and infraspinatus muscles, respectively.
- Cords:
    - From the lateral cord, C5, C6, and C7 supply the pectoralis major and minor muscles, via the lateral and medial pectoral nerves, as well as the coracobrachialis, brachialis and biceps brachii, via the musculocutaneous nerve. The musculocutaneous nerve provides sensation to the skin of the lateral forearm.
    - These three roots also form the lateral root of the median nerve that controls most forearm flexors, except flexor carpi ulnaris and the medial portion of flexor digitorum profundus, the first and second lumbricals and the muscles of the thenar eminence. This nerve also provides sensation to the palmar aspect of the thumb, index finger, middle finger, and the lateral aspect of the fourth digit.
    - The posterior cord has five outputs:
        - From C5 and C6, the upper and lower subscapular nerves supply the upper and lower portions of the subscapularis. The lower subscapular nerve also innervates teres major.
        - Additionally, from C5 and C6, the axillary nerve supplies the motor function of the deltoid and teres minor, as well as the sensory aspect of the overlying skin, the superior lateral cutaneous nerve of the arm, and the skin of the lateral shoulder and arm.
        - Thoracodorsal nerve originates from C6, C7, and C8, and innervates the latissimus dorsi.
        - Lastly, the radial nerve, which derives from all five roots of the brachial plexus, is responsible for the motor innervation of the triceps brachii, the supinator, the anconeus, the brachioradialis, and all extensor muscles of the forearm. It also supplies the skin of the posterior arm, posterior hand, and the webbing between the thumb and index finger.
    - The medial cord also produces five outputs, all of which derive from C8 and T1:
        - The medial pectoral nerve, which supplies pectoralis minor and part of pectoralis major.

        - The medial root of the median nerve.
        - The medial cutaneous nerves of the arm and forearm are purely sensory and supply the medial skin of arm and forearm, respectively.
        - Lastly, is the ulnar nerve, which provides motor control to the flexor carpi ulnaris, the medial belly of flexor digitorum profundus, the two medial lumbricals, the interossei, and the muscles of the hypothenar eminence. It also provides sensation to the fifth digit, the hypothenar eminence and the medial half of the fourth finger.
    - Branches:
        - Musculocutaneous nerve.
        - Axillary nerve.
        - Radial nerve.
        - Median nerve.
        - Ulnar nerve.

Embryology

Spinal nerves originate from the sclerotome, more specifically, the central region of the anterior-half-sclerotome. This portion of the sclerotome is also responsible for the endoneurium and perineurium of the spinal nerves. This origin stands in contrast to the posterior-half-sclerotome that forms the bone and cartilage of the vertebral column. The posterior half also acts to block the outgrowth of neural crest cells and spinal nerve axons, thus leading to the segmentation of the spinal nerves.

Blood Supply and Lymphatics

The roots of spinal nerves receive vascular supply from the radicular arteries. These arteries pass through the intervertebral foramina with the nerve roots. The first six vessels most commonly arise from the vertebral arteries or the ascending cervical branch of the thyrocervical trunk. In either case, the spinal branches of these two vessels anastomose, so if either source has a blockage, blood flow is maintained. The radicular arteries of the seventh and eighth segments always form from branches of the costocervical trunk.

Physiologic Variants

There are numerous variants of cervical nerves and their derivatives that can present in a patient. Specifically, the cervical nerves that contribute to the cervical and brachial plexuses vary between individuals to a significant degree. Moreover, the terminal nerves that branch off the cervical and brachial plexuses may vary in contribution and location of origin.

Occiput

The occipital bone is a cranial dermal bone and the main bone of the occiput (back and lower part of the skull). It is trapezoidal in shape and curved on itself like a shallow dish. The occipital bone overlies the occipital lobes of the cerebrum. At the base of skull in the occipital bone, there is a large oval opening called the foramen magnum, which allows the passage of the spinal cord.

Like the other cranial bones, it is classed as a flat bone. Due to its many attachments and features, the occipital bone is described in terms of separate parts. From its front to the back is the basilar part, also called the basioccipital, at the sides of the foramen magnum are the lateral parts, also called the exoccipitals, and the back is named as the squamous part.

The basilar part is a thick, somewhat quadrilateral piece in front of the foramen magnum and directed towards the pharynx.

The squamous part is the curved, expanded plate behind the foramen magnum and is the largest part of the occipital bone.

Structure

The occipital bone, like the other seven cranial bones, has outer and inner layers (also called plates or tables) of cortical bone tissue between which is the cancellous bone tissue known in the cranial bones as diploë. The bone is especially thick at the ridges, protuberances, condyles, and anterior part of the basilar part; in the inferior cerebellar fossae it is thin, semitransparent, and without diploë.

Outer Surface

Near the middle of the outer surface of the squamous part of the occipital (the largest part) there is a prominence – the external occipital protuberance. The highest point of this is called the inion. From the inion, along the midline of the squamous part until the foramen magnum, runs a ridge – the external occipital crest (also called the medial nuchal line) and this gives attachment to the nuchal ligament.

Running across the outside of the occipital bone are three curved lines and one line (the medial line) that runs down to the foramen magnum. These are known as the nuchal lines which give attachment to various ligaments and muscles. They are named as the highest, superior and inferior nuchal lines. The inferior nuchal line runs across the midpoint of the *median* nuchal line. The area above the highest nuchal line is termed the occipital plane and the area below this line is termed the nuchal plane.

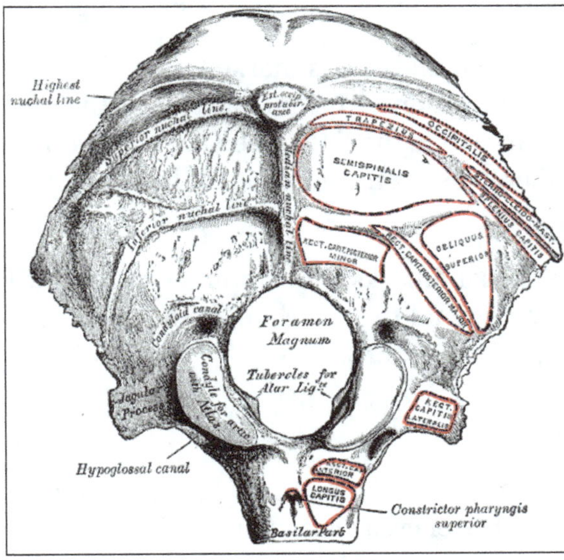

Outer surface of occipital bone.

Inner Surface

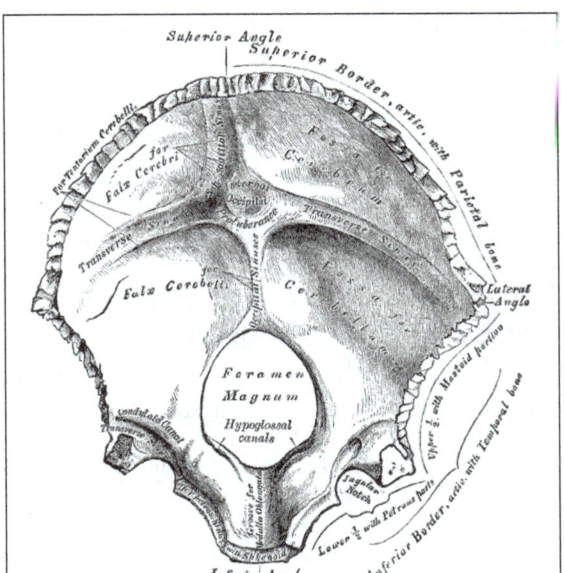

Inner surface of occipital bone.

The inner surface of the occipital bone forms the base of the posterior cranial fossa. The foramen magnum is a large hole situated in the middle, with the clivus, a smooth part of the occipital bone travelling upwards in front of it. The median internal occipital crest travels behind it to the internal occipital protuberance, and serves as a point of attachment to the falx cerebri.

To the sides of the foramen sitting at the junction between the lateral and base of the occipital bone are the hypoglossal canals. Further out, at each junction between the occipital and petrous portion of the temporal bone lies a jugular foramen.

The inner surface of the occipital bone is marked by dividing lines as shallow ridges, that form four fossae or depressions. The lines are called the cruciform (cross-shaped) eminence. At the midpoint where the lines intersect a raised part is formed called the internal occipital protuberance. From each side of this eminence runs a groove for the transverse sinuses.

There are two midline skull landmarks at the foramen magnum. The basion is the most anterior point of the opening and the opisthion is the point on the opposite posterior part. The basion lines up with the dens.

Foramen Magnum

The foramen magnum is a large oval foramen longest front to back; it is wider behind than in front where it is encroached upon by the occipital condyles. The clivus, a smooth bony section, travels upwards on the front surface of the foramen, and the median internal occipital crest travels behind it.

Through the foramen passes the medulla oblongata and its membranes, the accessory nerves, the vertebral arteries, the anterior and posterior spinal arteries, and the tectorial membrane and alar ligaments.

Angles

- The superior angle of the occipital bone articulates with the occipital angles of the parietal bones and, in the fetal skull, corresponds in position with the posterior fontanelle.
- The lateral angles are situated at the extremities of the groove for the transverse sinuses: each is received into the interval between the mastoid angle of the parietal bone, and the mastoid portion of the temporal bone.
- The inferior angle is fused with the body of the sphenoid bone.

Borders

- The superior borders extend from the superior to the lateral angles: they are deeply serrated for articulation with the occipital borders of the parietals, and form by this union the lambdoidal suture.
- The inferior borders extend from the lateral angles to the inferior angle; the upper half of each articulates with the mastoid portion of the corresponding temporal, the lower half with the petrous part of the same bone.

These two portions of the inferior border are separated from one another by the jugular process, the notch on the anterior surface of which forms the posterior part of the jugular foramen.

Sutures

- The lambdoid suture joins the occipital bone to the parietal bones.
- The occipitomastoid suture joins the occipital bone and mastoid portion of the temporal bone.
- The sphenobasilar suture joins the basilar part of the occipital bone and the back of the sphenoid bone body.
- The petrous-basilar suture joins the side edge of the basilar part of the occipital bone to the petrous-part of the temporal bone.

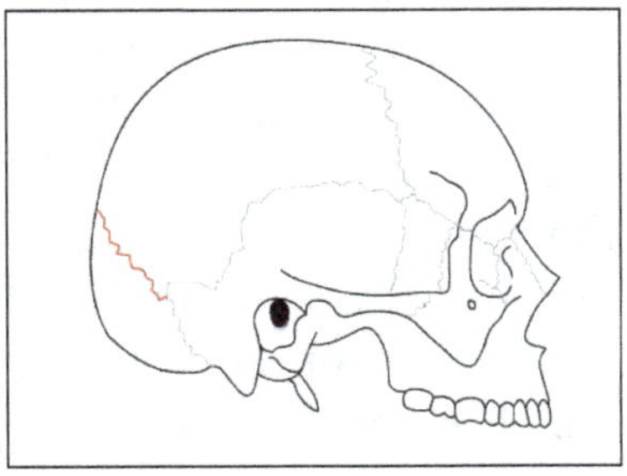

Lambdoid suture.

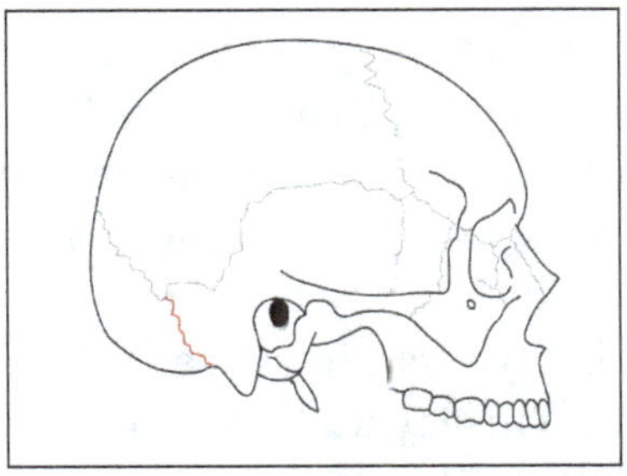

Occipitomastoid suture.

Scalp

The scalp is the anatomical area bordered by the human face at the front, and by the neck at the sides and back.

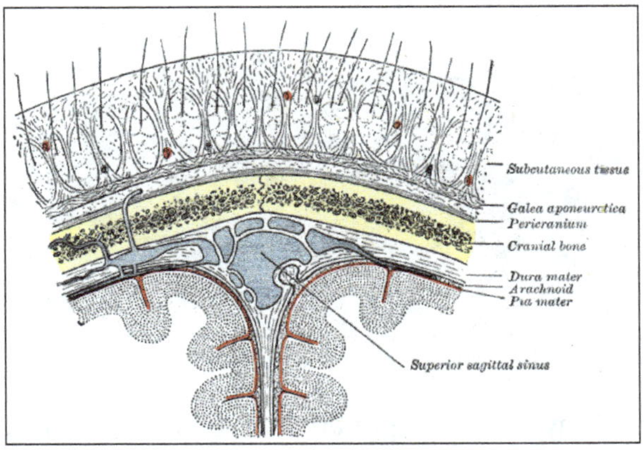

Diagrammatic section of scalp.

Structure

The scalp is usually described as having five layers, which can conveniently be remembered as a mnemonic:

- S: The skin on the head from which head hair grows. It contains numerous sebaceous glands and hair follicles.
- C: Connective tissue. A dense subcutaneous layer of fat and fibrous tissue that lies beneath the skin, containing the nerves and vessels of the scalp.
- A: The aponeurosis called epicranial aponeurosis (or galea aponeurotica) is the next layer. It is a tough layer of dense fibrous tissue which runs from the frontalis muscle anteriorly to the occipitalis posteriorly.
- L: The loose areolar connective tissue layer provides an easy plane of separation between the upper three layers and the pericranium. In scalping the scalp is torn off through this layer. It also provides a plane of access in craniofacial surgery and neurosurgery. This layer is sometimes referred to as the "danger zone" because of the ease by which infectious agents can spread through it to emissary veins which then drain into the cranium. The loose areolar tissue in this layer is made up of random collagen I bundles, collagen III. It will also be rich in glycosaminoglycans (GAGs) and will be constituted of more matrix than fibers. This layer allows the more superficial layers of the scalp to shift about in relation to the pericranium.
- P: The pericranium is the periosteum of the skull bones and provides nutrition to the bone and the capacity for repair. It may be lifted from the bone to allow removal of bone windows (craniotomy).

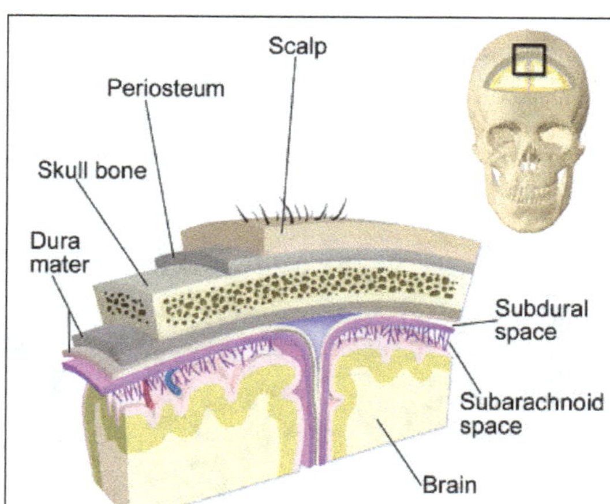

Illustration of the scalp and meninges.

The clinically important layer is the aponeurosis. Scalp lacerations through this layer mean that the "anchoring" of the superficial layers is lost and gaping of the wound occurs which would require suturing. This can be achieved with simple or vertical mattress sutures using a non-absorbable material, which are subsequently removed at around days 7-10.

Blood Supply

The blood supply of the scalp is via five pairs of arteries, three from the external carotid and two from the internal carotid:

- Internal carotid:
    - The supratrochlear artery to the midline forehead. The supratrochlear artery is a branch of the ophthalmic branch of the internal carotid artery.
    - The supraorbital artery to the lateral forehead and scalp as far up as the vertex. The supraorbital artery is a branch of the ophthalmic branch of the internal carotid artery.
- External carotid:
    - The superficial temporal artery gives off frontal and parietal branches to supply much of the scalp.
    - The occipital artery which runs posteriorly to supply much of the posterior aspect of the scalp.
    - The posterior auricular artery, a branch of the external carotid artery, ascends. behind the auricle to supply the scalp above and behind the auricle.

Because the walls of the blood vessels are firmly attached to the fibrous tissue of the superficial fascial layer, cut ends of vessels here do not readily retract; even a small scalp wound may bleed profusely.

Venous Drainage

The veins of the scalp accompany the arteries and thus have similar names, e.g. Supratrochlear and supraorbital veins, which unite at the medial angle of the eye, and form the angular vein, which further continues as the facial vein. The superficial temporal vein descends in front of the tragus, enters the parotid gland, and then joins the maxillary vein to form the retromandibular vein. The anterior part of it unites with the facial vein to form the common facial vein, which drains into jugular vein, and ultimately to the subclavian vein. The occipital vein terminates to the sub-occipital plexus.

There are other veins, like the emissary vein and frontal diploic vein, which also contribute to the venous drainage.

Nerve Supply

Innervation is the connection of nerves to the scalp: the sensory and motor nerves innervating the scalp. The scalp is innervated by the following:

- Supratrochlear nerve and the supraorbital nerve from the ophthalmic division of the trigeminal nerve.
- Greater occipital nerve (C2) posteriorly up to the vertex.
- Lesser occipital nerve (C2) behind the ear.
- Zygomaticotemporal nerve from the maxillary division of the trigeminal nerve supplying the hairless temple.
- Auriculotemporal nerve from the mandibular division of the trigeminal nerve.

The innervation of scalp can be remembered using the mnemonic, "Z-GLASS" for, Zygomaticotemporal nerve, Greater occipital nerve, Lesser occipital nerve, Auriculotemporal nerve, Supratrochlear nerve and Supraorbital nerve.

Lymphatic Drainage

Lymphatic channels from the posterior half of the scalp drain to occipital and posterior auricular nodes. Lymphatic channels from the anterior half drain to the parotid nodes. The lymph eventually reaches the submandibular and deep cervical nodes.

Ear

The ear is the organ of hearing and, in mammals, balance. In mammals, the ear is usually described as having three parts—the outer ear, the middle ear and the inner ear. The outer ear consists of the pinna and the ear canal. Since the outer ear is the only visible portion of the ear in most animals, the word "ear" often refers to the external part alone. The middle ear includes the tympanic cavity and the three ossicles. The inner ear sits in the bony labyrinth, and contains structures which are key to several senses: the semicircular canals, which enable balance and eye tracking when moving; the utricle and saccule, which enable balance when stationary; and the cochlea, which enables hearing. The ears of vertebrates are placed somewhat symmetrically on either side of the head, an arrangement that aids sound localisation.

The ear develops from the first pharyngeal pouch and six small swellings that develop in the early embryo called otic placodes, which are derived from ectoderm. The ear may be affected by disease, including infection and traumatic damage. Diseases of the ear may lead to hearing loss, tinnitus and balance disorders such as vertigo, although many of these conditions may also be affected by damage to the brain or neural pathways leading from the ear.

The ear has been adorned by earrings and other jewelry in numerous cultures for thousands of years, and has been subjected to surgical and cosmetic alterations.

Structure

The human ear consists of three parts—the outer ear, middle ear and inner ear. The ear canal of the outer ear is separated from the air-filled tympanic cavity of the middle ear by the eardrum. The middle ear contains the three small bones—the ossicles—involved in the transmission of sound, and is connected to the throat at the nasopharynx, via the pharyngeal opening of the Eustachian tube. The inner ear contains the otolith organs—the utricle and saccule—and the semicircular canals belonging to the vestibular system, as well as the cochlea of the auditory system.

Outer Ear

The outer ear is the external portion of the ear and includes the fleshy visible pinna (also called the auricle), the ear canal, and the outer layer of the eardrum (also called the tympanic membrane).

The pinna consists of the curving outer rim called the helix, the inner curved rim called the antihelix, and opens into the ear canal. The tragus protrudes and partially obscures the ear canal, as does the facing antitragus. The hollow region in front of the ear canal is called the concha. The ear canal stretches for about 1 inch (2.5 cm). The first part of the canal is surrounded by cartilage, while the second part near the eardrum is surrounded by bone. This bony part is known as the auditory bulla and is formed by the tympanic part of the temporal bone. The skin surrounding the ear canal contains ceruminous and sebaceous glands that produce protective ear wax. The ear canal ends at the external surface of the eardrum.

Two sets of muscles are associated with the outer ear: the intrinsic and extrinsic muscles. In some mammals, these muscles can adjust the direction of the pinna. In humans, these muscles have little or no effect. The ear muscles are supplied by the facial nerve, which also supplies sensation to the skin of the ear itself, as well as to the external ear cavity. The great auricular nerve, auricular nerve, auriculotemporal nerve, and lesser and greater occipital nerves of the cervical plexus all supply sensation to parts of the outer ear and the surrounding skin.

The pinna consists of a single piece of elastic cartilage with a complicated relief on its inner surface and a fairly smooth configuration on its posterior surface. A tubercle, known as Darwin's tubercle, is sometimes present, lying in the descending part of the helix and corresponding to the ear-tip of mammals. The earlobe consists of areola and adipose tissue. The symmetrical arrangement of the two ears allows for the localisation of sound. The brain accomplishes this by comparing arrival-times and intensities from each ear, in circuits located in the superior olivary complex and the trapezoid bodies which are connected via pathways to both ears.

Middle Ear

The middle ear lies between the outer ear and the inner ear. It consists of an air-filled cavity called the tympanic cavity and includes the three ossicles and their attaching ligaments; the auditory tube; and the round and oval windows. The ossicles are three small bones that function together to receive, amplify, and transmit the sound from the eardrum to the inner ear. The ossicles are the malleus (hammer), incus (anvil), and the stapes (stirrup). The stapes is the smallest named bone in the body. The middle ear also connects to the upper throat at the nasopharynx via the pharyngeal opening of the Eustachian tube.

The three ossicles transmit sound from the outer ear to the inner ear. The malleus receives vibrations from sound pressure on the eardrum, where it is connected at its longest part (the manubrium or handle) by a ligament. It transmits vibrations to the incus, which in turn transmits the vibrations to the small stapes bone. The wide base of the stapes rests on the oval window. As the stapes vibrates, vibrations are transmitted through the oval window, causing movement of fluid within the cochlea.

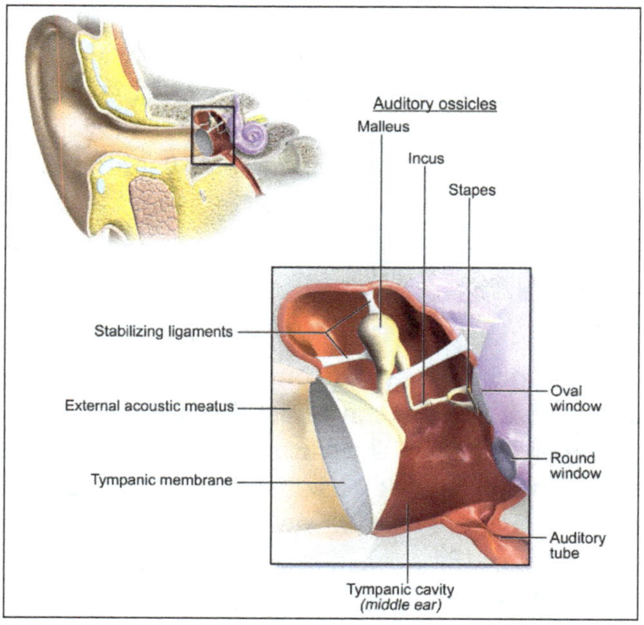

The middle ear.

The round window allows for the fluid within the inner ear to move. As the stapes pushes the secondary tympanic membrane, fluid in the inner ear moves and pushes the membrane of the round window out by a corresponding amount into the middle ear. The ossicles help amplify sound waves by nearly 15–20 times.

Inner Ear

The inner ear sits within the temporal bone in a complex cavity called the bony labyrinth. A central area known as the vestibule contains two small fluid-filled recesses, the utricle and saccule. These connect to the semicircular canals and the cochlea. There are three semicircular canals angled at right angles to each other which are responsible for dynamic balance. The cochlea is a spiral shell-shaped organ responsible for the sense of hearing. These structures together create the membranous labyrinth.

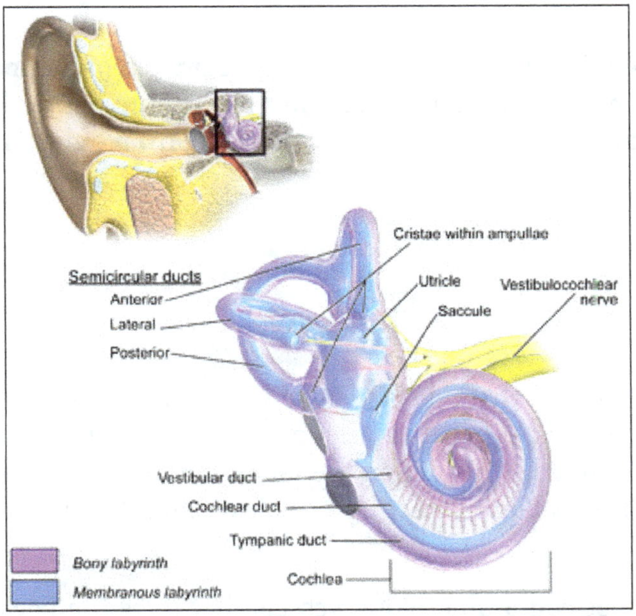

The outer ear receives sound, transmitted through the ossicles of the middle ear to the inner ear, where it is converted to a nervous signal in the cochlear and transmitted along the vestibulocochlear nerve.

The bony labyrinth refers to the bony compartment which contains the membranous labyrinth, contained within the temporal bone. The inner ear structurally begins at the oval window, which receives vibrations from the incus of the middle ear. Vibrations are transmitted into the inner ear into a fluid called endolymph, which fills the membranous labyrinth. The endolymph is situated in two vestibules, the utricle and saccule, and eventually transmits to the cochlea, a spiral-shaped structure. The cochlea consists of three fluid-filled spaces: the vestibular duct, the cochlear duct, and the tympanic duct. Hair cells responsible for transduction—changing mechanical changes into electrical stimuli are present in the organ of Corti in the cochlea.

Blood Supply

The blood supply of the ear differs according to each part of the ear.

The outer ear is supplied by a number of arteries. The posterior auricular artery provides the majority of the blood supply. The anterior auricular arteries provide some supply to the outer rim of the ear and scalp behind it. The posterior auricular artery is a direct branch of the external carotid artery, and the anterior auricular arteries are branches from the superficial temporal artery. The occipital artery also plays a role.

The middle ear is supplied by the mastoid branch of either the occipital or posterior auricular arteries and the deep auricular artery, a branch of the maxillary artery. Other arteries which are present but play a smaller role include branches of the middle meningeal artery, ascending pharyngeal artery, internal carotid artery, and the artery of the pterygoid canal.

The inner ear is supplied by the anterior tympanic branch of the maxillary artery; the stylomastoid branch of the posterior auricular artery; the petrosal branch of middle meningeal artery; and the labyrinthine artery, arising from either the anterior inferior cerebellar artery or the basilar artery.

Eye

The human eye is an organ that reacts to light and allows vision. Rod and cone cells in the retina allow conscious light perception and vision including color differentiation and the perception of depth. The human eye can differentiate between about 10 million colors and is possibly capable of detecting a single photon. The eye is part of the sensory nervous system.

- Choroid: Layer containing blood vessels that lines the back of the eye and is located between the retina (the inner light-sensitive layer) and the sclera (the outer white eye wall).
- Ciliary Body: Structure containing muscle and is located behind the iris, which focuses the lens.
- Cornea: The clear front window of the eye which transmits and focuses (i.e., sharpness or clarity) light into the eye. Corrective laser surgery reshapes the cornea, changing the focus.
- Fovea: The center of the macula which provides the sharp vision.
- Iris: The colored part of the eye which helps regulates the amount of light entering the eye. When there is bright light, the iris closes the pupil to let in less light. And when there is low light, the iris opens up the pupil to let in more light.
- Lens: Focuses light rays onto the retina. The lens is transparent, and can be replaced if necessary. Our lens deteriorates as we age, resulting in the need for reading glasses. Intraocular lenses are used to replace lenses clouded by cataracts.

- Macula: The area in the retina that contains special light-sensitive cells. In the macula these light-sensitive cells allow us to see fine details clearly in the center of our visual field. The deterioration of the macula is a common condition as we get older (age related macular degeneration or ARMD).

- Optic Nerve: A bundle of more than a million nerve fibers carrying visual messages from the retina to the brain. (In order to see, we must have light and our eyes must be connected to the brain). Your brain actually controls what you see, since it combines images. The retina sees images upside down but the brain turns images right side up. This reversal of the images that we see is much like a mirror in a camera. Glaucoma is one of the most common eye conditions related to optic nerve damage.

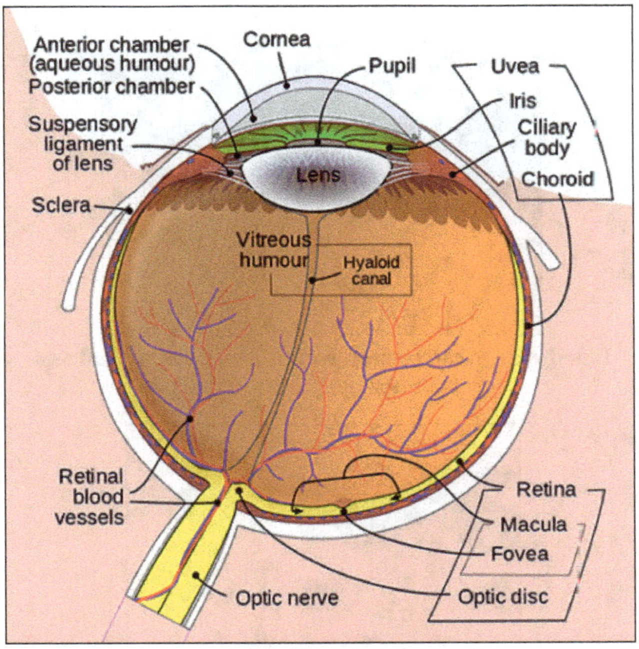

Anatomy of the Eye.

- Pupil: The dark center opening in the middle of the iris. The pupil changes size to adjust for the amount of light available (smaller for bright light and larger for low light). This opening and closing of light into the eye is much like the aperture in most 35 mm cameras which lets in more or less light depending upon the conditions.

- Retina: The nerve layer lining the back of the eye. The retina senses light and creates electrical impulses that are sent through the optic nerve to the brain.

- Sclera: The white outer coat of the eye, surrounding the iris.

- Vitreous Humor: The, clear, gelatinous substance filling the central cavity of the eye.

Structure

The eye is not shaped like a perfect sphere, rather it is a fused two-piece unit, composed of the anterior segment and the posterior segment. The anterior segment is made up of the cornea, iris and lens. The cornea is transparent and more curved, and is linked to the larger posterior segment, composed of the vitreous, retina, choroid and the outer white shell called the sclera. The cornea is typically about 11.5 mm (0.3 in) in diameter, and 0.5 mm (500 μm) in thickness near its center. The posterior chamber constitutes the remaining five-sixths; its diameter is typically about 24 mm. The cornea and sclera are connected by an area termed the limbus. The iris is the pigmented circular structure concentrically surrounding the center of the eye, the pupil, which appears to be black. The size of the pupil, which controls the amount of light entering the eye, is adjusted by the iris' dilator and sphincter muscles.

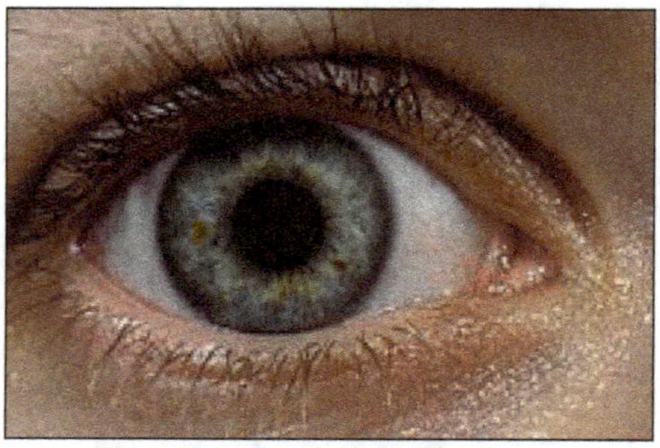

Blood vessels can be seen within the sclera, as well as a strong limbal ring around the iris.

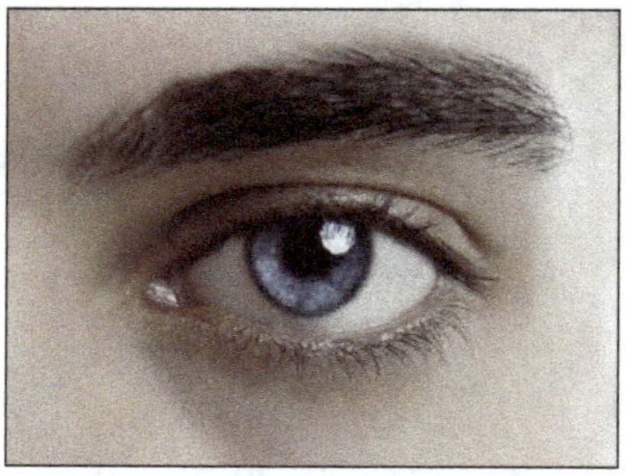

The outer parts of the eye.

Light energy enters the eye through the cornea, through the pupil and then through the lens. The lens shape is changed for near focus (accommodation) and is controlled by the ciliary muscle. Photons of light falling on the light-sensitive cells of the retina (photoreceptor cones and rods) are converted into electrical signals that are transmitted to the brain by the optic nerve and interpreted as sight and vision.

Size

The size of the eye differs among adults by only one or two millimetres. The eyeball is generally less tall than it is wide. The sagittal vertical (height) of a human adult eye is approximately 23.7 mm, the transverse horizontal diameter (width) is 24.2 mm and the axial anteroposterior size (depth) averages 22.0–24.8 mm with no significant difference between sexes and age groups. Strong correlation has been found between the transverse diameter and the width of the orbit (r = 0.88). The typical adult eye has an anterior to posterior diameter of 24 millimetres, and a volume of six cubic centimetres (0.4 cu. in.).

The eyeball grows rapidly, increasing from about 16–17 millimetres (about 0.65 inch) at birth to 22.5–23 mm (approx. 0.89 in) by three years of age. By age 12, the eye attains its full size.

Components

The eye is made up of three coats, or layers, enclosing various anatomical structures. The outermost layer, known as the fibrous tunic, is composed of the cornea and sclera , which provide shape to the eye and support the deeper structures. The middle layer, known as the vascular tunic or uvea, consists of the choroid, ciliary body, pigmented epithelium and iris. The innermost is the retina, which gets its oxygenation from the blood vessels of the choroid (posteriorly) as well as the retinal vessels (anteriorly).

The spaces of the eye are filled with the aqueous humour anteriorly, between the cornea and lens, and the vitreous body, a jelly-like substance, behind the lens, filling the entire posterior cavity. The aqueous humour is a clear watery fluid that is contained in two areas: the anterior chamber between the cornea and the iris, and the posterior chamber between the iris and the lens. The lens is suspended to the ciliary body by the suspensory ligament (Zonule of Zinn), made up of hundreds of fine transparent fibers which transmit muscular forces to change the shape of the lens for accommodation (focusing). The vitreous body is a clear substance composed of water and proteins, which give it a jelly-like and sticky composition.

Vision

Field of View

The approximate field of view of an individual human eye (measured from the fixation point, i.e., the point at which one's gaze is directed) varies by facial anatomy, but is typically 30° superior (up, limited by the brow), 45° nasal (limited by the nose), 70° inferior (down), and 100° temporal (towards the temple). For both eyes combined (binocular) visual field is 135° vertical and 200° horizontal. It is an area of 4.17 steradians or 13700 square degrees for binocular vision. When viewed at large angles from the side, the iris and pupil may still be visible by the viewer, indicating the person has peripheral vision possible at that angle.

About 15° temporal and 1.5° below the horizontal is the blind spot created by the optic nerve nasally, which is roughly 7.5° high and 5.5° wide.

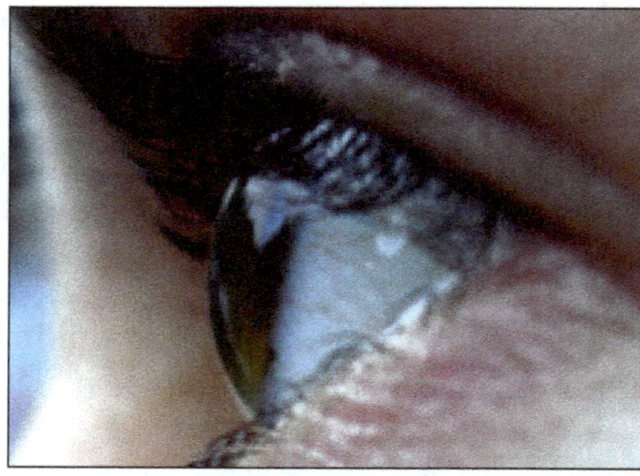

Side-view of the human eye, viewed approximately 90° temporal, illustrating how the iris and pupil appear rotated towards the viewer due to the optical properties of the cornea and the aqueous humor.

Dynamic Range

The retina has a static contrast ratio of around 100:1 (about 6.5 f-stops). As soon as the eye moves rapidly to acquire a target (saccades), it re-adjusts its exposure by adjusting the iris, which adjusts the size of the pupil. Initial dark adaptation takes place in approximately four seconds of profound, uninterrupted darkness; full adaptation through adjustments in retinal rod photoreceptors is 80% complete in thirty minutes. The process is nonlinear and multifaceted, so an interruption by light exposure requires restarting the dark adaptation process over again. Full adaptation is dependent on good blood flow; thus dark adaptation may be hampered by retinal disease, poor vascular circulation and high altitude exposure.

The human eye can detect a luminance range of 10^{14}, or one hundred trillion (100,000,000,000,000) (about 46.5 f-stops), from 10^{-6} cd/m^2, or one millionth (0.000001) of a candela per square meter to 10^8 cd/m^2 or one hundred million (100,000,000) candelas per square meter. This range does not include looking at the midday sun (10^9 cd/m^2) or lightning discharge.

At the low end of the range is the absolute threshold of vision for a steady light across a wide field of view, about 10^{-6} cd/m2 (0.000001 candela per square meter). The upper end of the range is given in terms of normal visual performance as 10^8 cd/m^2 (100,000,000 or one hundred million candelas per square meter).

The eye includes a lens similar to lenses found in optical instruments such as cameras and the same physics principles can be applied. The pupil of the human eye is its aperture; the iris is the diaphragm that serves as the aperture stop. Refraction in the cornea causes the effective aperture (the entrance pupil) to differ slightly from the physical pupil diameter. The entrance pupil is typically about 4 mm in diameter, although it can range from 2 mm (*f*/8.3) in a brightly lit place to 8 mm (*f*/2.1) in the dark. The latter value decreases slowly with age; older people's eyes sometimes dilate to not more than 5–6mm in the dark, and may be as small as 1mm in the light.

Eye Movement

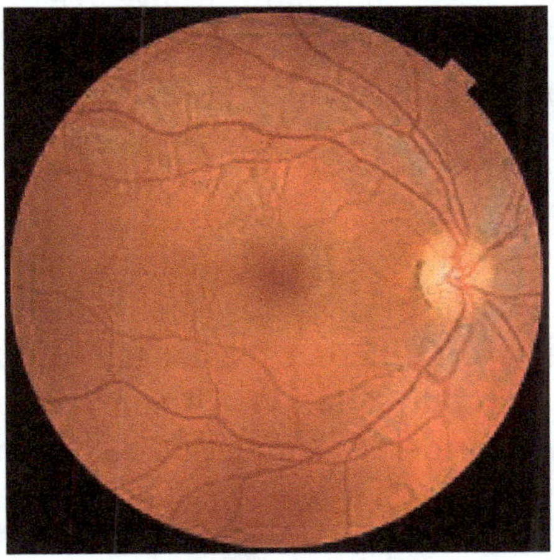

The light circle is the optic disc where the optic nerve exits the retina.

The visual system in the human brain is too slow to process information if images are slipping across the retina at more than a few degrees per second. Thus, to be able to see while moving, the brain must compensate for the motion of the head by turning the eyes. Frontal-eyed animals have a small area of the retina with very high visual acuity, the fovea centralis. It covers about 2 degrees of visual angle in people. To get a clear view of the world, the brain must turn the eyes so that the image of the object of regard falls on the fovea. Any failure to make eye movements correctly can lead to serious visual degradation.

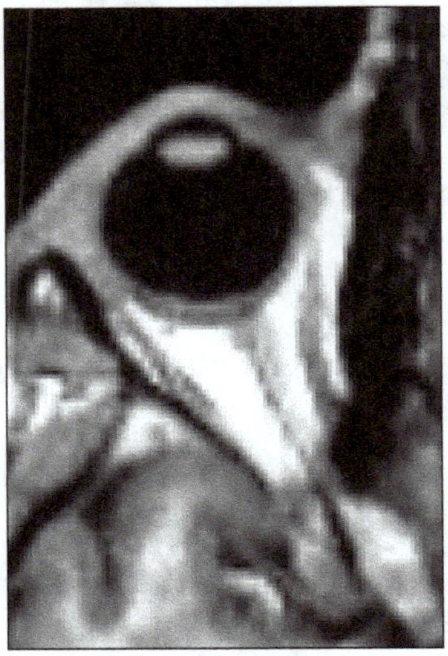

MRI scan of human eye.

Having two eyes allows the brain to determine the depth and distance of an object, called stereovision, and gives the sense of three-dimensionality to the vision. Both eyes must point accurately enough that the object of regard falls on corresponding points of the two retinas to stimulate stereovision; otherwise, double vision might occur. Some persons with congenitally crossed eyes tend to ignore one eye's vision, thus do not suffer double vision, and do not have stereovision. The movements of the eye are controlled by six muscles attached to each eye, and allow the eye to elevate, depress, converge, diverge and roll. These muscles are both controlled voluntarily and involuntarily to track objects and correct for simultaneous head movements.

Extraocular Muscles

Each eye has six muscles that control its movements: the lateral rectus, the medial rectus, the inferior rectus, the superior rectus, the inferior oblique, and the superior oblique. When the muscles exert different tensions, a torque is exerted on the globe that causes it to turn, in almost pure rotation, with only about one millimeter of translation. Thus, the eye can be considered as undergoing rotations about a single point in the center of the eye.

Rapid Eye Movement

Rapid eye movement, REM, typically refers to the sleep stage during which the most vivid dreams occur. During this stage, the eyes move rapidly.

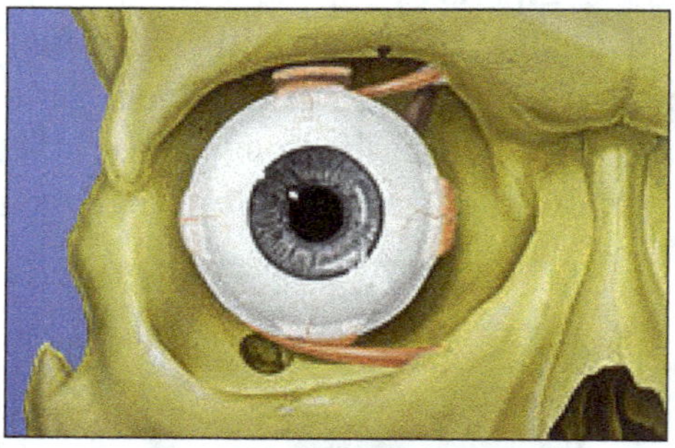

Normal anatomy of the human eye and orbit, anterior view.

Saccades

Saccades are quick, simultaneous movements of both eyes in the same direction controlled by the frontal lobe of the brain.

Fixational Eye Movements

Even when looking intently at a single spot, the eyes drift around. This ensures that individual photosensitive cells are continually stimulated in different degrees. Without changing input, these cells would otherwise stop generating output.

A 2019 study showed that small continuous and unconscious eye movements are needed to enable the detection of contrast, and for vision in general. Eye movements include drift, ocular tremor, and microsaccades. Some irregular drifts, movements smaller than a saccade and larger than a microsaccade, subtend up to one tenth of a degree. Researchers vary in their definition of Microsaccades by amplitude, Martin Rolfs states that 'the majority of microsaccades observed in a variety of tasks have amplitudes smaller than 30 min-arc'.

Vestibulo-ocular Reflexes

The vestibulo-ocular reflex is a reflex eye movement that stabilizes images on the retina during head movement by producing an eye movement in the direction opposite to head movement in response to neural input from the vestibular system of the inner ear, thus maintaining the image in the center of the visual field. For example, when the head moves to the right, the eyes move to the left. This applies for head movements up and down, left and right, and tilt to the right and left, all of which give input to the ocular muscles to maintain visual stability.

Smooth Pursuit Movement

Eyes can also follow a moving object around. This tracking is less accurate than the vestibulo-ocular reflex, as it requires the brain to process incoming visual information and supply feedback. Following an object moving at constant speed is relatively easy, though the eyes will often make saccadic jerks to keep up. The smooth pursuit movement can move the eye at up to 100°/s in adult humans.

It is more difficult to visually estimate speed in low light conditions or while moving, unless there is another point of reference for determining speed.

Optokinetic Reflex

The Optokinetic reflex (or optokinetic nystagmus) stabilizes the image on the retina through visual feedback. It is induced when the entire visual scene drifts across the retina, eliciting eye rotation in the same direction and at a velocity that minimizes the motion of the image on the retina. When the gaze direction deviates too far from the forward heading, a compensatory saccade is induced to reset the gaze to the centre of the visual field.

For example, when looking out of the window at a moving train, the eyes can focus on a moving train for a short moment (by stabilizing it on the retina), until the train moves out of the field of vision. At this point, the eye is moved back to the point where it first saw the train (through a saccade).

Near Response

The adjustment to close-range vision involves three processes to focus an image on the retina.

Vergence Movement

When a creature with binocular vision looks at an object, the eyes must rotate around a vertical axis so that the projection of the image is in the centre of the retina in both eyes. To look at a nearby object, the eyes rotate 'towards each other' (convergence), while for an object farther away they rotate 'away from each other' (divergence).

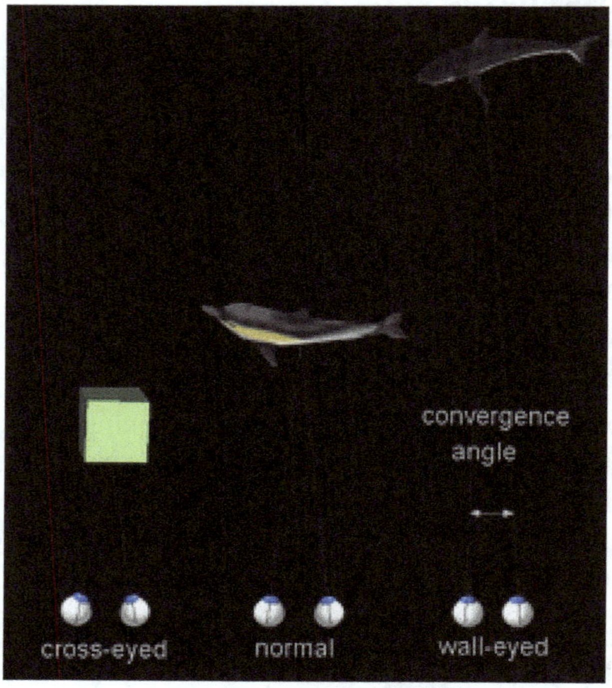

The two eyes converge to point to the same object.

Pupil Constriction

Lenses cannot refract light rays at their edges as well as they can closer to the center. The image produced by any lens is therefore somewhat blurry around the edges (spherical aberration). It can be minimized by screening out peripheral light rays and looking only at the better-focused center. In the eye, the pupil serves this purpose by constricting while the eye is focused on nearby objects. Small apertures also give an increase in depth of field, allowing a broader range of "in focus" vision. In this way the pupil has a dual purpose for near vision: to reduce spherical aberration and increase depth of field.

Accommodation of the Lens

Changing the curvature of the lens is carried out by the ciliary muscles surrounding the lens; this process is known as "accommodation". Accommodation narrows the inner diameter of the ciliary body, which actually relaxes the fibers of the suspensory ligament attached to the periphery of the lens, and also allows the lens to relax into a more convex, or globular, shape. A more convex lens refracts light more strongly and focuses divergent light rays from near objects onto the retina, allowing closer objects to be brought into better focus.

Nose

Nose is the prominent structure between the eyes that serves as the entrance to the respiratory tract and contains the olfactory organ. It provides air for respiration, serves the sense of smell, conditions the air by filtering, warming, and moistening it, and cleans itself of foreign debris extracted from inhalations.

The nose has two cavities, separated from one another by a wall of cartilage called the septum. The external openings are known as nares or nostrils. The roof of the mouth and the floor of the nose are formed by the palatine bone, the mouth part of which is commonly called the hard palate; a flap of tissue, the soft palate, extends back into the nasopharynx, the nasal portion of the throat, and during swallowing is pressed upward, thus closing off the nasopharynx so that food is not lodged in the back of the nose.

The shape of the nasal cavity is complex. The forward section, within and above each nostril, is called the vestibule. Behind the vestibule and along each outer wall are three elevations, running generally from front to rear. Each elevation, called a nasal concha or turbinate, hangs over an air passage. Beside and above the uppermost concha is the olfactory region of the nasal cavity. The rest of the cavity is the respiratory portion. The respiratory area is lined with a moist mucous membrane with fine hairlike projections known as cilia, which serve to collect debris. Mucus from cells in the membrane wall also helps to trap particles of dust, carbon, soot, and bacteria. Sinus cavities are located in the bony skull on both sides of the nose.

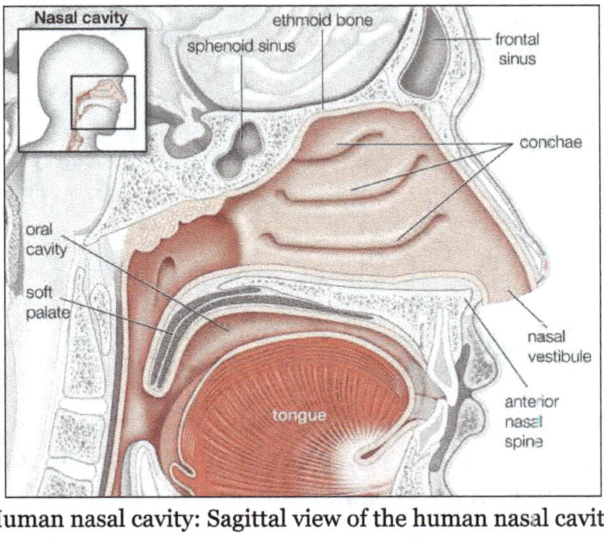

Human nasal cavity: Sagittal view of the human nasal cavity.

In the olfactory (smelling) portion of the nose, most of the lining is mucous membrane. A small segment of the lining contains the nerve cells that are the actual sensory organs. Fibres, called dendrites, which project from the nerve cells into the nasal cavity, are covered only by a thin layer of moisture. The moisture dissolves microscopic particles that the air has carried into the nose from odour-emitting substances, and the particles dissolved in the fluid stimulate the olfactory nerve cells chemically.

Structure

Several bones and cartilages make up the bony-cartilaginous framework of the nose, and the internal structure. The nose is also made up of types of soft tissue such as skin, epithelia, mucous membrane, muscles, nerves, and blood. In the skin there are sebaceous glands, and in the mucous membrane there are nasal glands. The bones and cartilages provide strong protection for the internal structures of the nose. There are several muscles that are involved in movements of the nose. The arrangement of the cartilages allows flexibility through muscle control to enable airflow to be modified.

Bones

The bony structure of the nose is provided by the maxilla, frontal bone, and a number of smaller bones.

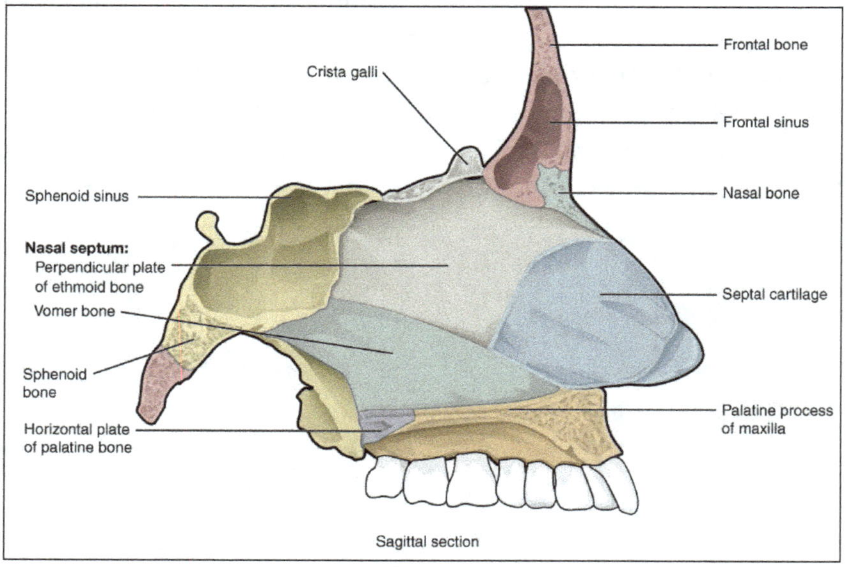

Bones of the nose and septal cartilage.

The topmost bony part of the nose is formed by the nasal part of the frontal bone, which lies between the brow ridges, and ends in a serrated nasal notch. A left and a right nasal bone join with the nasal part of the frontal bone at either side; and these at the side with the small lacrimal bones and the frontal process of each maxilla. The internal roof of the nasal cavity is composed of the horizontal, perforated cribriform plate of the ethmoid bone through which pass sensory fibres of the olfactory nerve. Below and behind the cribriform plate, sloping down at an angle, is the face of the sphenoid bone.

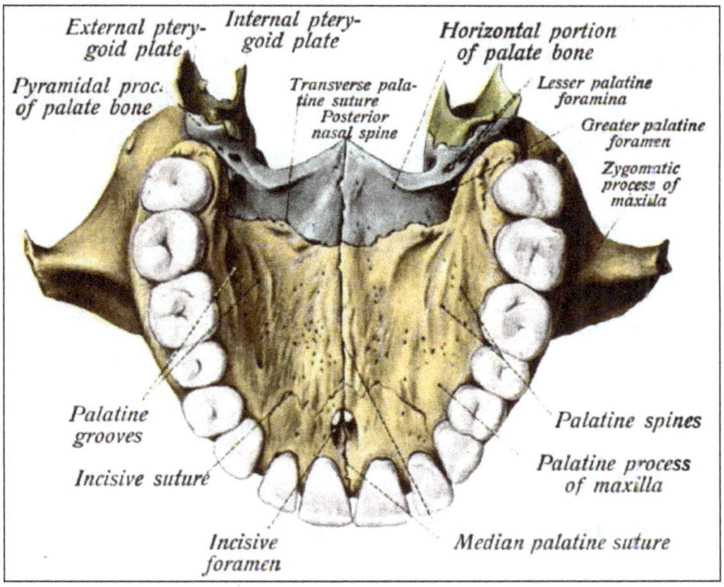

Roof of the mouth showing position of palatine bones making up the floor of the nose, and forming the posterior nasal spine for the attachment of the musculus uvulae.

The wall separating the two cavities of the nose, the nasal septum, is made up of bone inside and cartilage closer to the tip of the nose. The bony part is formed by the perpendicular plate of the ethmoid bone at the top, and the vomer bone that below. The floor of the nose is made up of the incisive bone and the horizontal plates of the palatine bones, and this makes up the hard palate of the roof of the mouth. The two horizontal plates join together at the midline and form the posterior nasal spine that gives attachment to the musculus uvulae in the uvula.

The two maxilla bones join at the base of the nose at the lower nasal midline between the nostrils, and at the top of the philtrum to form the anterior nasal spine. This thin projection of bone holds the cartilaginous center of the nose. It is also an important cephalometric landmark.

Cartilages

The nasal cartilages are the septal, lateral, major alar, and minor alar cartilages. The major and minor cartilages are also known as the greater and lesser alar cartilages. There is a narrow strip of cartilage called the vomeronasal cartilage that lies between the vomer and the septal cartilage.

The septal nasal cartilage, extends from the nasal bones in the midline, to the bony part of the septum in the midline, posteriorly. It then passes along the floor of the nasal cavity. The septum is quadrangular–the upper half is attached to the two lateral nasal cartilages which are fused to the dorsal septum in the midline. The septum is laterally attached, with loose ligaments, to the bony margin of the anterior nasal aperture, while the inferior ends of the lateral cartilages are free (unattached). The three or four minor alar cartilages are adjacent to the lateral cartilages, held in the connective tissue membrane, that connects the lateral cartilages to the frontal process of the maxilla.

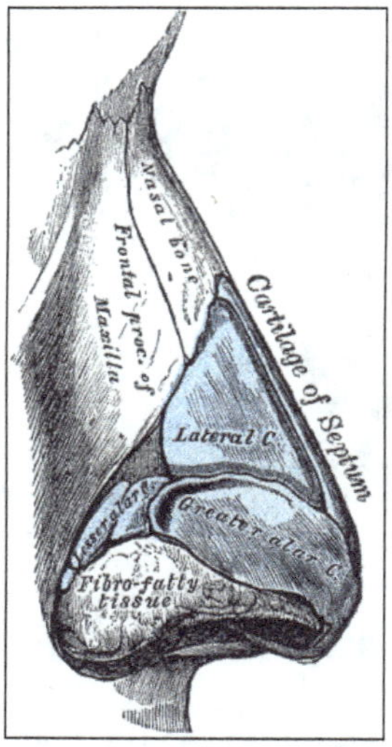

Nasal cartilages.

The nasal bones in the upper part of the nose are joined together by the midline internasal suture. They join with the septal cartilage at a junction known as the **rhinion**. The rhinion is the midpoint of the internasal suture at the join with the cartilage, and from the rhinion to the apex, or tip, the framework is of cartilage.

The major alar cartilages are thin, U-shaped plates of cartilage on each side of the nose that form the lateral and medial walls of the vestibule, known as the medial and lateral crura. The medial crura are attached to the septal cartilage, forming fleshy parts at the front of the nostrils on each side of the septum, called the *medial crural footpods*. The medial crura meet at the midline below the end of the septum to form the columella and lobule. The lobule contains the tip of the nose and its base contains the nostrils. At the peaks of the folds of the medial crura, they form the alar domes the tip-defining points of the nose, separated by a notch. They then fold outwards, above and to the side of the nostrils forming the lateral crura. The major alar cartilages are freely moveable and can respond to muscles to either open or constrict the nostrils.

There is a reinforcing structure known as the nasal scroll that resists internal collapse from airflow pressure generated by normal breathing. This structure is formed by the junction between the lateral and major cartilages. Their edges interlock by one scrolling upwards and one scrolling inwards.

Muscles

The muscles of the nose are a subgroup of the muscles of facial expression. They are involved in respiration and facial expression. The muscles of the nose include the procerus, nasalis, depressor septi nasalis, levator labii superioris alaeque nasalis, and the orbicularis oris of the mouth. As are all of the facial muscles, the muscles of the nose are innervated by the facial nerve and its branches. Although each muscle is independent the muscles of the nose form a continuous layer with connections between all the components of the muscles and ligaments, in the nasal part of a superficial muscular aponeurotic system. It is continuous from the nasofrontal process to the nasal tip. It divides at level of the nasal valve into superficial and deep layers, each layer having medial and lateral components.

The procerus muscle produces wrinkling over the bridge of the nose, and is active in concentration and frowning. It is a prime target for Botox procedures in the forehead to remove the lines between the eyes.

The nasalis muscle consists of two main parts: a transverse part called the compressor naris, and an alar part termed the dilator naris. The compressor naris muscle compresses the nostrils and may completely close them. The alar part, the dilator naris mainly consists of the dilator naris posterior, and a much smaller dilator naris anterior, and this muscle flares the nostrils. The dilator naris helps to form the upper ridge of the philtrum. The anterior, and the posterior dilator naris, (the alar part of the nasalis muscle), give support to the nasal valves.

The depressor septi muscle may sometimes be absent or rudimentary. The depressor septi pulls the columella, the septum, and the tip of the nose downwards. At the start of inspiration this muscle tenses the nasal septum and with the dilator naris widens the nostrils.

The levator labii superioris alaeque nasi, divides into a medial and a lateral slip. The medial slip blends into the perichondrium of the major alar cartilage and its overlying skin. The lateral slip blends at the side of the upper lip with the levator labii superioris, and with the orbicularis oris. The lateral slip raises the upper lip and deepens and increases the curve above the nasolabial furrow. The medial slip pulls the lateral crus upwards and modifies the curve of the furrow around the alae, and dilates the nostrils.

Soft Tissue

The skin of the nose varies in thickness along its length. From the glabella to the bridge (the nasofrontal angle) the skin is thick, fairly flexible, and mobile. It tapers to the bridge where it is thinnest and least flexible as it is closest to the underlying bone. From the bridge until the tip of the nose the skin is thin. The tip is covered in skin that is as thick as the top section, and has many large sebaceous glands. The thickness of the skin varies but is still separated from the underlying bones and cartilage by four layers – a superficial fatty layer; a fibromuscular layer continued from the SMAS; a deep fatty layer, and the periosteum.

Other areas of soft tissue are found where there is no support from cartilage - these include an area around the sides of the septum - the paraseptal area; an area around the lateral cartliages; an area at the top of the nostril, and an area in the alae.

External Nose

The nasal root is the top of the nose that attaches the nose to the forehead. The nasal root is above the bridge and below the glabella, forming an indentation known as the nasion at the frontonasal suture where the frontal bone meets the nasal bones.

The nasal dorsum also known as the *nasal ridge* is the border between the root and the tip of the nose which in profile can be variously shaped.

The ala of the nose (ala nasi, wing of the nose) is the lower lateral surface of the external nose, shaped by the alar cartilage and covered in dense connective tissue. The alae flare out to form a rounded eminence around the nostril.

Sexual dimorphism is evident in the larger nose of the male. This is due to the increased testosterone that thickens the brow ridge and the bridge of the nose making it wider.

Nasal Cavity

The nasal cavity is the large internal space of the nose, and is in two parts – the nasal vestibule and the nasal cavity proper. The nasal vestibule is the frontmost part of the nasal cavity, enclosed by cartilages. The vestibule is lined with skin, hair follicles, and a large number of sebaceous glands. A mucous ridge known as the limen nasi separates the vestibule from the rest of the nasal cavity and marks the change from the skin of the vestibule to the respiratory epithelium of the rest of the nasal cavity. This area is also known as a mucocutaneous junction that has a dense microvasculature.

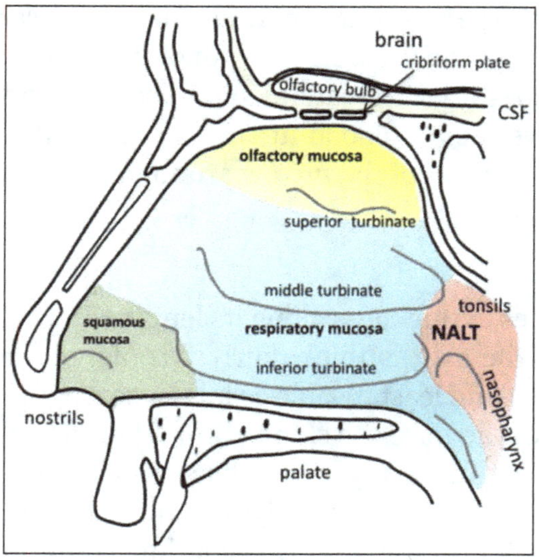

Anatomy of the nasal cavity. Nasal-associated lymphoid tissue labelled NALT.

The nasal cavity is divided into two cavities, by the nasal septum, and each is accessed by an external nostril. The division into two cavities enables the functioning of the nasal cycle that slows down the air conditioning process. At the back of the nasal cavity there are two openings, called choanae (also *posterior nostrils*), that give entrance to the nasopharynx, and rest of the respiratory tract.

On the outer wall of each cavity are three shell-like bones called conchae, arranged as superior, middle and inferior nasal conchae. Below each concha is a corresponding superior, middle, and inferior nasal meatus, or passage. Sometimes when the superior concha is narrow, a fourth supreme nasal concha is present situated above and sharing the space with the superior concha. The term concha refers to the actual bone - when covered by soft tissue and mucosa, and functioning, a concha is termed a turbinate.

Most of the nasal cavity and paranasal sinuses is lined with respiratory epithelium as nasal mucosa. In the roof of each cavity is an area of specialised olfactory epithelium. This region is about 5 square cm, covering the superior concha, the cribriform plate, and the nasal septum.

Nasal Valves

There is a nasal valve area in the cavity responsible for providing resistance to the flow of air. This enables an increased time for warming and moistening the air. This area is of two nasal valves. The internal nasal valve is the narrowest part of the airway in the middle third of the cavity. The larger external nasal valve is located in the alar wall. The angle between the septum and the sidewall needs to be sufficient for unobstructed airflow.

Borders

Each half of the nasal cavity is bounded by a roof, floor, medial wall (the septum), and lateral wall. The middle part of the roof of the nasal cavity is composed of the horizontal, perforated cribriform plate of the ethmoid bone, through which pass sensory fibres of the olfactory nerve into the cranial cavity.

Nose Shape

The shape of the nose varies widely due to differences in the nasal bone shapes and formation of the bridge of the nose. Some nose shapes were classified for surgeries by Eden Warwick in *Nasology* 1848:

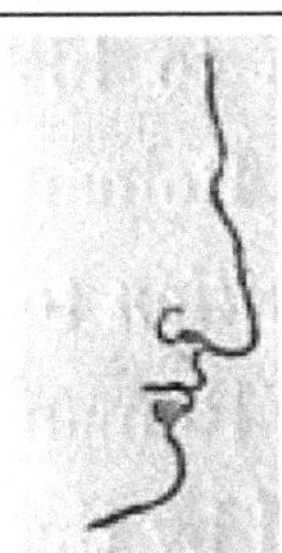		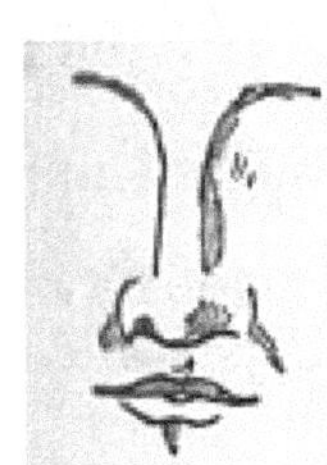
Class I. The aquiline nose.	Class II. The Grecian nose.	Class III. The African nose.

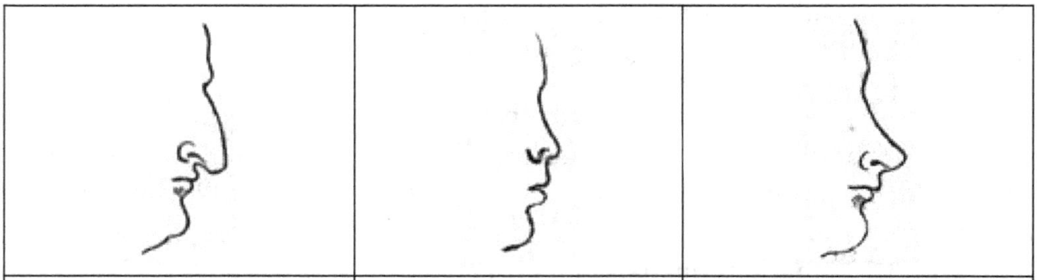

Class IV. The hawk nose.	Class V. The snub nose.	Class VI. The celestial nose.

Other terms are used to describe the shape of the nose, and can also include a reference to the nasal index used to classify ethnicity: *Leptorrhine* describes a long, narrow nose. *Hyperleptorrhine* is a very long, narrow nose with a nasal index of 40 to 55. The nasal index is the ratio of the width of the nose to the nasal height. Some of the types used in *Nasology* are also used by Topinard. He took narrow shapes as in 1-5 to be European; medium width shape 6 were "yellow races"; broad nose shape 7 were African, and broad nose shape 8 were taken to be Melanesian and Australian Aboriginal.

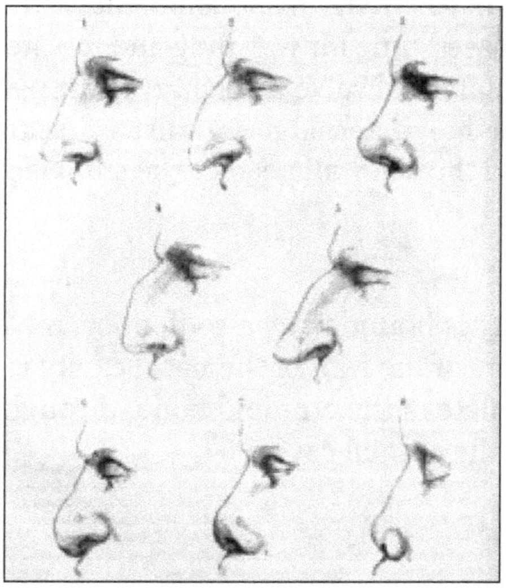

Nose shapes used in Topinard's nasal index.

Some deformities of the nose are named, such as the pug nose and the saddle nose. The pug nose is characterised by excess tissue from the apex that is out of proportion to the rest of the nose. A low and underdeveloped nasal bridge may also be evident. A saddle nose is mostly associated with trauma to the nose but can be caused by other conditions including leprosy.

Werner syndrome, a condition that causes the appearance of premature aging, causes a "bird-like" appearance due to pinching of the nose.

Down syndrome commonly presents a small nose with a flattened nasal bridge. This can be due to the absence of one or both nasal bones, shortened nasal bones, or nasal bones that have not fused in the midline.

Lymphatic Drainage

From different areas of the nose superficial lymphatic vessels run with the veins, and deep lymphatic vessels travel with the arteries. Lymph drains from the anterior half of the nasal cavity, including both the medial and lateral walls, to join that of the external nasal skin to drain into the submandibular lymph nodes. The rest of the nasal cavity and paranasal sinuses all drain to the upper deep cervical lymph nodes, either directly or through the retropharyngeal lymph nodes. The back of the nasal floor probably drains to the parotid lymph nodes.

Development of the Nose

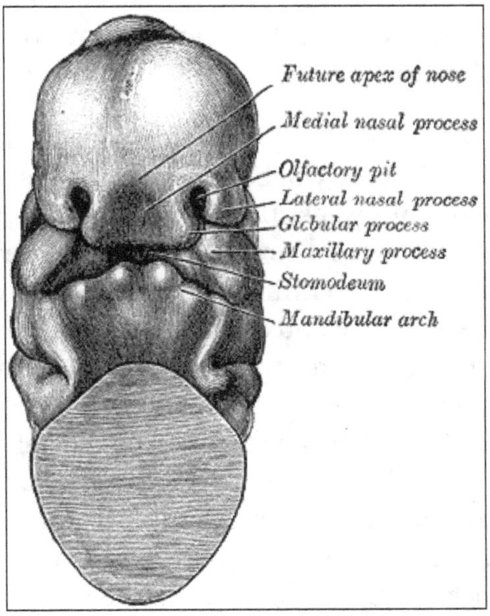

The developing head at about four weeks. The olfactory pit can be seen at the top.

In the early development of the embryo, neural crest cells migrate to form the mesenchymal tissue as ectomesenchyme of the pharyngeal arches. By the end of the fourth week, the first pair of pharyngeal arches form five facial prominences or processes - an unpaired frontonasal process, paired mandibular processes and paired maxillary processes. The nose is largely formed by the fusion of these five facial prominences. The frontonasal process gives rise to the bridge of the nose. The medial nasal processes provide the crest and the tip of the nose, and the lateral nasal processes form the alae or sides of the nose. The frontonasal process is a proliferation of mesenchyme in front of the brain vesicles, and makes up the upper border of the stomadeum.

During the fifth week the maxillary processes increase in size and at the same time the ectoderm of the frontonasal process becomes thickened at its sides and also increases in size, forming the nasal placodes. The nasal placodes are also known as the olfactory placodes. This development is induced by the ventral part of the forebrain. In the sixth week the ectoderm in each nasal placode invaginates to form an indented oval-shaped pit, which forms a surrounding raised ridge of tissue. Each nasal pit forms a division between the ridges, into a lateral nasal process on the outer edge, and a medial nasal process on the inner edge.

In the sixth week the nasal pits deepen as they penetrate into the underlying mesenchyme. At this time, the medial nasal processes migrate towards each other and fuse forming the primordium of the bridge of the nose and the septum. The migration is helped by the increased growth of the maxillary prominences medially, which compresses the medial nasal processes towards the midline. Their merging takes place at the surface, and also at a deeper level. The merge forms the intermaxillary segment, and this is continuous with the rostral part of the nasal septum. The tips of the maxillary processes also grow and fuse with the intermaxillary process. The intermaxillary process gives rise to the philtrum of the upper lip.

At the end of the sixth week the nasal pits have deepened further and they fuse to make a large ectodermal nasal sac. This sac will be above and to the back of the intermaxillary process. Leading into the seventh week, the nasal sac floor and posterior wall grow to form a thickened plate-like ectodermal structure called the nasal fin. The nasal fin separates the sac from the oral cavity. Within the fin, vacuoles develop that fuse with the nasal sac. This enlarges the nasal sac and at the same time thins the fin to a membrane - the oronasal membrane that separates the nasal pits from the oral cavity. During the seventh week the oronasal membrane ruptures and disintegrates to form an opening - the single primitive choana. The intermaxillary segment extends posteriorly to form the primary palate which makes up the floor of the nasal cavity. During the eighth and ninth weeks a pair of thin extensions form from the medial walls of the maxillary process. These extensions are called the palatine shelves that form the secondary palate. The secondary palate will endochondrally ossify to form the hard palate - the end-stage floor of the nasal cavity. During this time ectoderm and mesoderm of the frontonasal process produce the midline septum. The septum grows down from the roof of the nasal cavity and fuses with the developing palates along the midline. The septum divides the nasal cavity into two nasal passages opening into the pharynx through the definitive choanae.

At ten weeks, the cells differentiate into muscle, cartilage, and bone. Problems at this stage of development can cause birth defects such as choanal atresia (absent or closed passage), facial clefts and nasal dysplasia (faulty or incomplete development) or extremely rarely polyrrhinia the formation of a duplicate nose.

Normal development is critical because the newborn infant breathes through the nose for the first six weeks, and any nasal blockage will need emergency treatment to clear.

Development of the Paranasal Sinuses

The four pairs of paranasal sinuses - the maxillary, ethmoid, sphenoid, and frontal, develop from the nasal cavity as invaginations extending into their named bones. Two pairs of sinuses form during prenatal development and two pairs form after birth. The maxillary sinuses are the first to appear during the fetal third month. They slowly expand within the maxillary bones and continue to expand throughout childhood. The maxillary sinuses form as invaginations from the nasal sac. The ethmoid sinuses appear in the fetal fifth month as invaginations of the middle meatus. The ethmoid sinuses do not grow into the ethmoid bone and do not completely develop until puberty.

The sphenoid sinuses are extensions of the ethmoid sinuses into the sphenoid bones. They begin to develop around two years of age, and continue to enlarge during childhood.

The frontal sinuses only develop in the fifth or sixth year of childhood, and continue expanding throughout adolescence. Each frontal sinus is made up of two independent spaces that develop from two different sources; one from the expansion of ethmoid sinuses into frontal bone, and the other develops from invagination. They never coalesce so drain independently.

Function

Respiration

The nose is the first organ of the upper respiratory tract in the respiratory system. Its main respiratory function is the supply and conditioning, by warming, moisturising and filtering of particulates of inhaled air. Nasal hair in the nostrils traps large particles preventing their entry into the lungs.

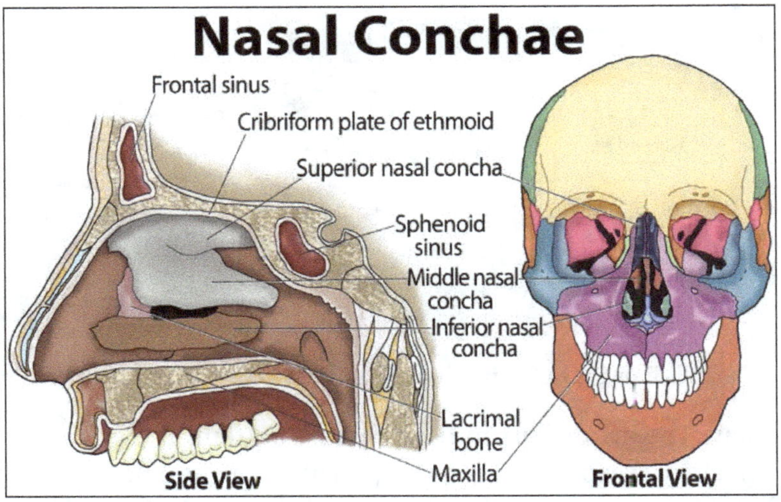

Air enters the upper respiratory tract through the nose. The nasal conchae can be seen in this image.

The three positioned nasal conchae in each cavity provide four grooves as air passages, along which the air is circulated and moved to the nasopharynx. The internal structures and cavities, including the conchae and paranasal sinuses form an integrated system for the conditioning of the air breathed in through the nose. This functioning also includes the major role of the nasal mucosa, and the resulting conditioning of the air before it reaches the lungs is important in maintaining the internal environment and proper functioning of the lungs. The turbulence created by the conchae and meatuses optimises the warming, moistening, and filtering of the mucosa. A major protective role is thereby provided by these structures of the upper respiratory tract, in the passage of air to the more delicate structures of the lower respiratory tract.

The nose is also able to provide sense information as to the temperature of the air being breathed.

Sneezing is an important protective reflex initiated by irritation of the nasal mucosa to expel unwanted particles through the mouth and nose. Photic sneezing is a reflex brought on by different stimuli such as bright lights.

Sense of Smell

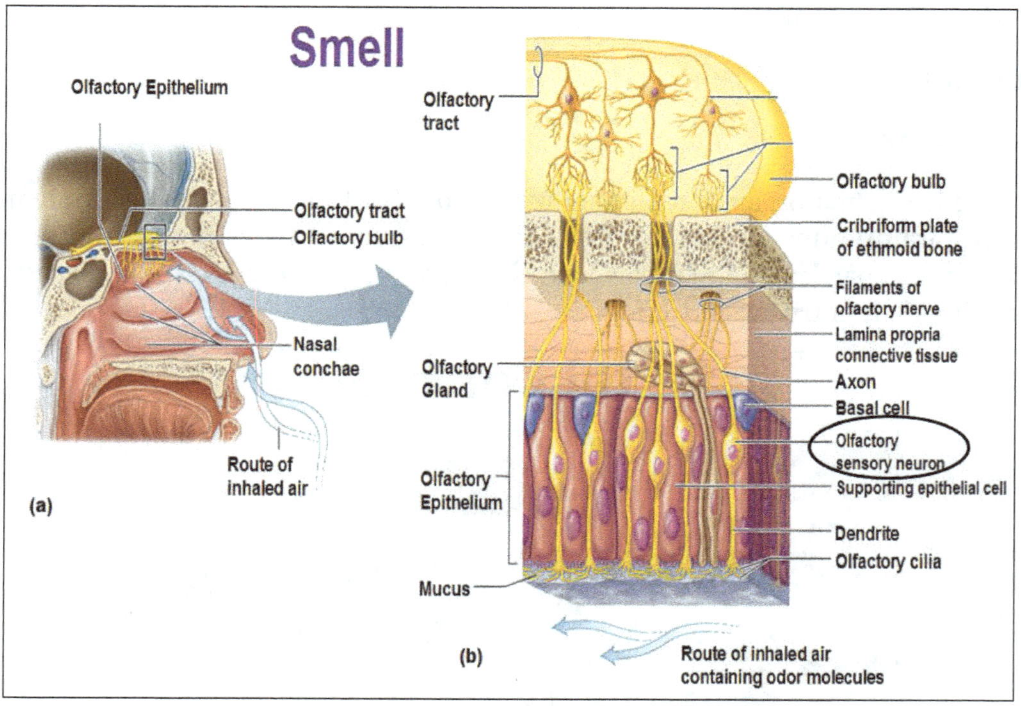

The cells lining the nose involved in smell (olfaction) are shown here.

The nose also plays the major part in the olfactory system. It contains an area of specialised cells, olfactory receptor neurons responsible for the sense of smell (olfaction). Olfactory mucosa in the upper nasal cavity, contains a type of nasal gland called olfactory glands or Bowman's glands which help in olfaction. The nasal conchae also help in olfaction function, by directing air-flow to the olfactory region.

Speech

Normal speech is produced with pressure from the lungs. This can be modified using airflow through the nose in a process called nasalisation. This involves the lowering of the soft palate to produce nasal vowels and consonants by allowing air to escape from both the nose and the mouth. Nasal airflow is also used to produce a variety of nasal clicks called click consonants. The paranasal sinuses act as sound chambers in vocal resonation.

Mouth

In human anatomy, the mouth is the first portion of the alimentary canal that receives food and produces saliva. The oral mucosa is the mucous membrane epithelium lining the inside of the mouth.

In addition to its primary role as the beginning of the digestive system, in humans the mouth also plays a significant role in communication. While primary aspects of the voice are produced in the throat, the tongue, lips, and jaw are also needed to produce the range of sounds included in human language.

The mouth consists of two regions, the vestibule and the oral cavity proper. The mouth, normally moist, is lined with a mucous membrane, and contains the teeth. The lips mark the transition from mucous membrane to skin, which covers most of the body.

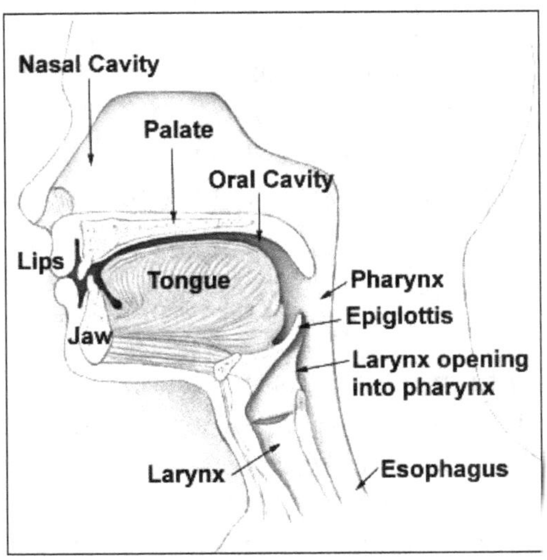

Structure

Oral Cavity

The mouth consists of 2 regions: the vestibule and the oral cavity proper. The vestibule is the area between the teeth, lips and cheeks. The oral cavity is bounded at the sides and in front by the alveolar process (containing the teeth) and at the back by the isthmus of the fauces. Its roof is formed by hard palate at the front, and a soft palate at the back. The uvula projects downwards from the middle of the soft palate at its back. The floor is formed by the mylohyoid muscles and is occupied mainly by the tongue. A mucous membrane – the oral mucosa, lines the sides and under surface of the tongue to the gums, lining the inner aspect of the jaw (mandible). It receives the secretions from the submandibular and sublingual salivary glands.

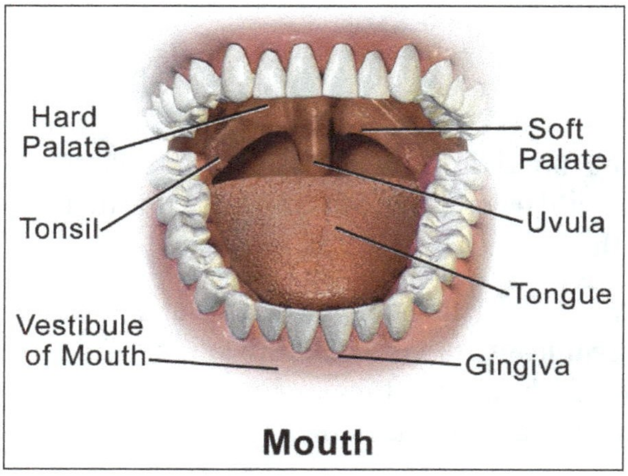

Anatomy of the mouth.

Orifice

While shut, the orifice of the mouth forms a line between the upper and lower lip. In facial expression, this *mouth line* is iconically shaped like an up-open parabola in a smile, and like a down-open parabola in a frown. A down-turned mouth means a mouth line forming a down-turned parabola, and when permanent can be normal. Also, a down-turned mouth can be part of the presentation of Prader-Willi syndrome.

Nerve Supply

The teeth and the periodontium (i.e. the tissues that support the teeth) are innervated by the maxillary and mandibular divisions of the trigeminal nerve. Maxillary (upper) teeth and their associated periodontal ligament are innervated by the superior alveolar nerves, branches of the maxillary division, termed the posterior superior alveolar nerve, anterior superior alveolar nerve, and the variably present middle superior alveolar nerve. These nerves form the superior dental plexus above the maxillary teeth. The mandibular (lower) teeth and their associated periodontal ligament are innervated by the inferior alveolar nerve, a branch of the mandibular division. This nerve runs inside the mandible, within the inferior alveolar canal below the mandibular teeth, giving off branches to all the lower teeth (inferior dental plexus). The oral mucosa of the gingiva (gums) on the facial (labial) aspect of the maxillary incisors, canines and premolar teeth is innervated by the superior labial branches of the infraorbital nerve. The posterior superior alveolar nerve supplies the gingiva on the facial aspect of the maxillary molar teeth. The gingiva on the palatal aspect of the maxillary teeth is innervated by the greater palatine nerve apart from in the incisor region, where it is the nasopalatine nerve (long sphenopalatine nerve). The gingiva of the lingual aspect of the mandibular teeth is innervated by the sublingual nerve, a branch of the lingual nerve. The gingiva on the facial aspect of the mandibular incisors and canines is innervated by the mental nerve, the continuation of the inferior alveolar nerve emerging from the mental foramen. The gingiva of the buccal (cheek) aspect of the mandibular molar teeth is innervated by the buccal nerve (long buccal nerve).

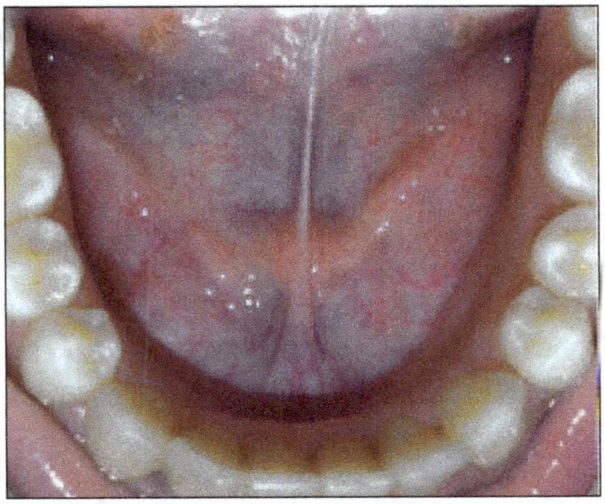

Floor of the mouth with Lingual frenum and sublingual fold.

Development

The philtrum is the vertical groves in the upper lip, formed where the nasomedial and maxillary processes meet during embryo development. When these processes fail to fuse fully, either a hare lip or cleft palate, (or both) can result.

The nasolabial folds are the deep creases of tissue that extend from the nose to the sides of the mouth. One of the first signs of age on the human face is the increase in prominence of the nasolabial folds.

Function

The mouth plays an important role in eating, drinking, breathing and speaking. Infants are born with a sucking reflex, by which they instinctively know to suck for nourishment using their lips and jaw. The mouth also helps in chewing and biting food.

For some disabled people, especially many disabled artists, who through illness, accident or congenital disability have lost dexterity, their mouths take the place of their hands, when typing, texting, writing, making drawings, paintings and other works of art by maneuvering brushes and other tools, in addition to the basic oral functions. Mouth painters hold the brush in their mouth or between their teeth and maneuver it with their tongue and cheek muscles, but mouth painting can be strenuous for neck and jaw muscles since the head has to perform the same back and forth movement as a hand does when painting.

A male mouth can hold, on average, 71.2 ml, while a female mouth holds 55.4 ml.

Jaw

Jaw is either of a pair of bones that form the framework of the mouth of vertebrate animals, usually containing teeth and including a movable lower jaw (mandible) and fixed upper jaw (maxilla). Jaws function by moving in opposition to each other and are used for biting, chewing, and the handling of food.

The mandible consists of a horizontal arch, which holds the teeth and contains blood vessels and nerves. Two vertical portions (rami) form movable hinge joints on either side of the head, articulating with the glenoid cavity of the temporal bone of the skull. The rami also provide attachment for muscles important in chewing. The centre front of the arch is thickened and buttressed to form a chin, a development unique to man and some of his recent ancestors; the great apes and other animals lack chins.

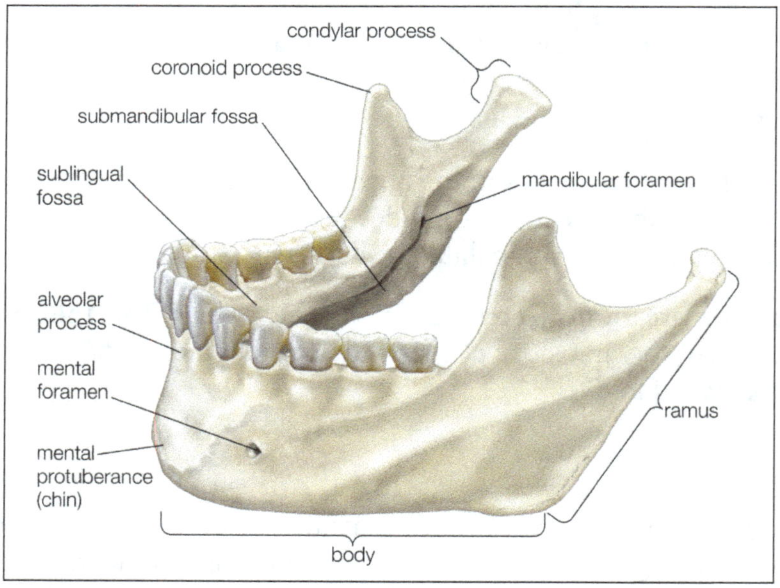

The mandible (lower jawbone).

The upper jaw is firmly attached to the nasal bones at the bridge of the nose; to the frontal, lacrimal, ethmoid, and zygomatic bones within the eye socket; to the palatine and sphenoid bones in the roof of the mouth; and at the side, by an extension, to the zygomatic bone (cheekbone), with which it forms the anterior portion of the zygomatic arch. The arched lower part of the maxilla contains the upper teeth. Inside the body of the bone is the large maxillary sinus.

In the human fetus and infant both the upper and lower jaws have two halves; these fuse at the midline a few months after birth.

Mandible

The mandible, lower jaw or jawbone is the largest, strongest and lowest bone in the human face. It forms the lower jaw and holds the lower teeth in place. The mandible sits beneath the maxilla. It is the only movable bone of the skull (discounting the ossicles of the middle ear).

The bone is formed in the fetus from a fusion of the left and right mandibular prominences, and the point where these sides join, the mandibular symphysis, is still visible as a faint ridge in the midline. Like other symphyses in the body, this is a midline articulation where the bones are joined by fibrocartilage, but this articulation fuses together in early childhood.

Structure

Components

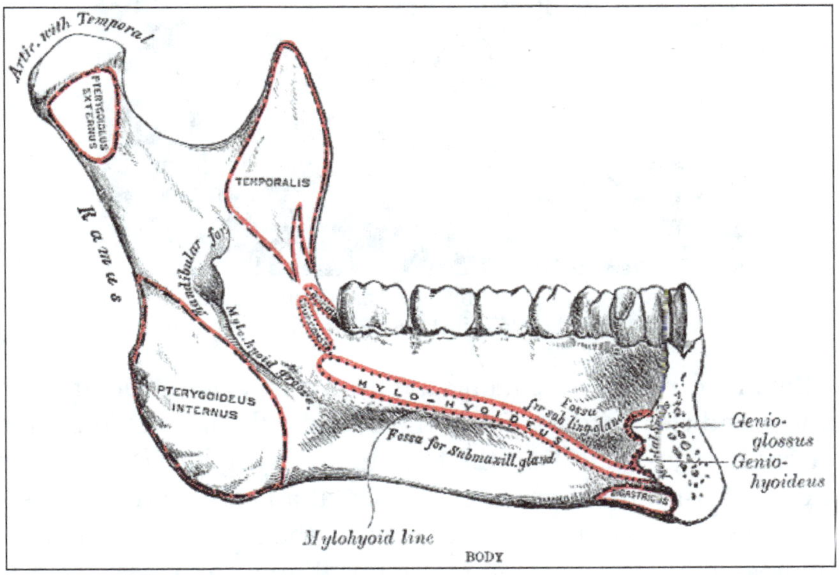

Mandible, medial surface, side view.

The mandible consists of:

- The body, found at the front.
- A ramus on the left and the right, the rami rise up from the body of the mandible and meet with the body at the angle of the mandible or the gonial angle.

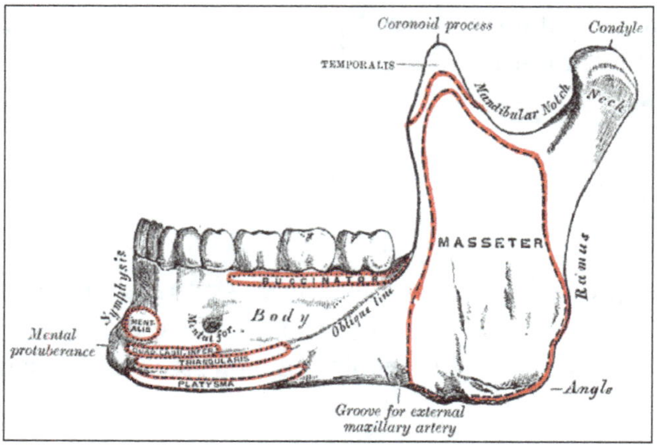

Mandible, lateral surface, side view.

Body

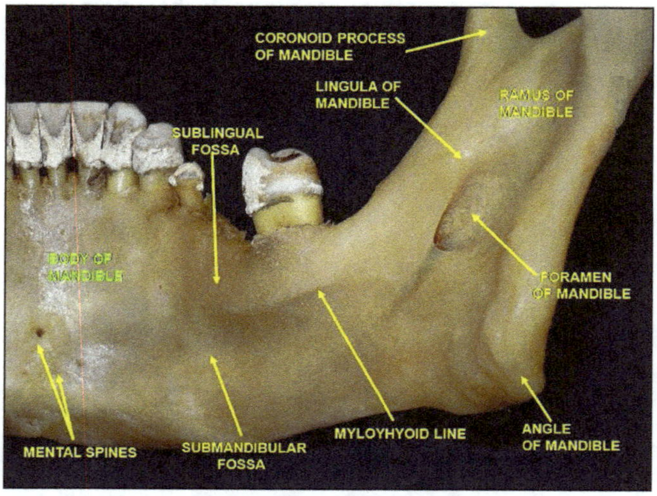

Body and ramus of the mandible: The mandibular foramen is labeled on the right. The lingula is just above the mandibular foramen.

The body of the mandible is curved, and the front part gives structure to the chin. It has two surfaces and two borders. From the outside, the mandible is marked in the midline by a faint ridge, indicating the mandibular symphysis, the line of junction of the two pieces of which the bone is composed at an early period of life. This ridge divides below and encloses a triangular eminence, the mental protuberance, the base of which is depressed in the center but raised on either side to form the mental tubercle. On either side of the symphysis, just below the incisor teeth, is a depression, the incisive fossa, which gives origin to the mentalis and a small portion of the orbicularis oris. Below the second premolar tooth, on either side, midway between the upper and lower borders of the body, is the mental foramen, for the passage of the mental vessels and nerve. Running backward and upward from each mental tubercle is a faint ridge, the oblique line, which is continuous with the anterior border of the ramus; it affords attachment to the depressor labii inferioris and depressor anguli oris; the platysma is attached below it.

From the inside, the mandible appears concave. Near the lower part of the symphysis is a pair of laterally placed spines, termed the mental spines, which give origin to the genioglossus. Immediately below these is a second pair of spines, or more frequently a median ridge or impression, for the origin of the geniohyoid. In some cases, the mental spines are fused to form a single eminence, in others they are absent and their position is indicated merely by an irregularity of the surface. Above the mental spines, a median foramen and furrow are sometimes seen; they mark the line of union of the halves of the bone. Below the mental spines, on either side of the middle line, is an oval depression for the attachment of the anterior belly of the digastric. Extending upward and backward on either side from the lower part of the symphysis is the mylohyoid line, which gives origin to the mylohyoid muscle; the posterior part of this line, near the alveolar margin, gives attachment to a small part of the constrictor pharyngis superior, and to the pterygomandibular raphe. Above the anterior part of this line is a smooth triangular area against which the sublingual gland rests, and below the hinder part, an oval fossa for the submandibular gland.

Borders:

- The superior or alveolar border, wider behind than in front, is hollowed into cavities, for the reception of the teeth; these cavities are sixteen in number and vary in depth and size according to the teeth which they contain. To the outer lip of the superior border, on either side, the buccinator is attached as far forward as the first molar tooth.
- The inferior border is rounded, longer than the superior, and thicker in front than behind; at the point where it joins the lower border of the ramus a shallow groove; for the facial artery, may be present.

Ramus

The ramus of the human mandible has four sides, two surfaces, four borders, and two processes. On the outside, the ramus is flat and marked by oblique ridges at its lower part; it gives attachment throughout nearly the whole of its extent to the masseter muscle.

On the inside at the center there is an oblique mandibular foramen, for the entrance of the inferior alveolar vessels and nerve. The margin of this opening is irregular; it presents in front a prominent ridge, surmounted by a sharp spine, the lingula of the mandible, which gives attachment to the sphenomandibular ligament; at its lower and back part is a notch from which the mylohyoid groove runs obliquely downward and forward, and lodges the mylohyoid vessels and nerve. Behind this groove is a rough surface, for the insertion of the medial pterygoid muscle. The mandibular canal runs obliquely downward and forward in the ramus, and then horizontally forward in the body, where it is placed under the alveoli and communicates with them by small openings. On arriving at the incisor teeth, it turns back to communicate with the mental foramen, giving off two small canals which run to the cavities containing the incisor teeth. In the posterior two-thirds of the bone the canal is situated nearer the internal surface of the mandible; and in the anterior third, nearer its external surface. It contains the inferior alveolar vessels and nerve, from which branches are distributed to the teeth.

Borders:

- The lower border of the ramus is thick, straight, and continuous with the inferior border of the body of the bone. At its junction with the posterior border is the angle of the mandible, which may be either inverted or everted and is marked by rough, oblique ridges on each side, for the attachment of the masseter laterally, and the medial pterygoid muscle medially; the stylomandibular ligament is attached to the angle between these muscles. The anterior border is thin above, thicker below, and continuous with the oblique line.
- The region where the lower border meets the posterior border is the angle of the mandible, often called the gonial angle.
- The posterior border is thick, smooth, rounded, and covered by the parotid gland. The upper border is thin, and is surmounted by two processes, the coronoid in front and the condyloid behind, separated by a deep concavity, the mandibular notch.

Processes:

- The coronoid process is a thin, triangular eminence, which is flattened from side to side and varies in shape and size.
- The condyloid process is thicker than the coronoid, and consists of two portions: the mandibular condyle, and the constricted portion which supports it, the neck.
- The mandibular notch, separating the two processes, is a deep semilunar depression and is crossed by the masseteric vessels and nerve.

Foramina

The mandible has two main holes (foramina), found on both its right and left sides:

- The mandibular foramen, is above the mandibular angle in the middle of each ramus.
- The mental foramen sits on either side of the mental protuberance (chin) on the body of mandible, usually inferior to the apices of the mandibular first and

second premolars. As mandibular growth proceeds in young children, the mental foramen alters in direction of its opening from anterior to posterosuperior. The mental foramen allows the entrance of the mental nerve and blood vessels into the mandibular canal.

Nerves

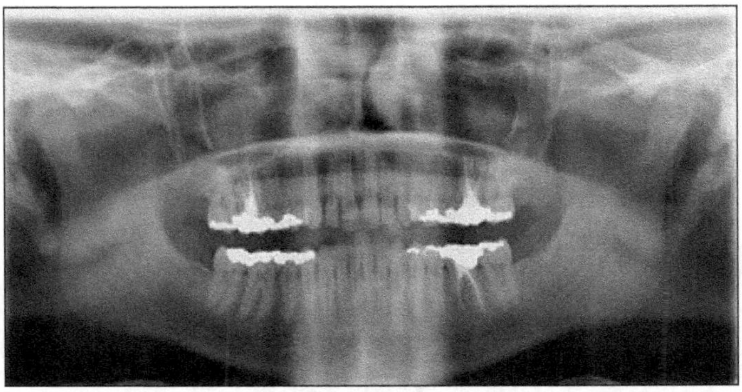

A panoramic radiograph reveals the mandible, including the heads and necks of the mandibular condyles, the coronoid processes of the mandible, as well as the nasal antrum and the maxillary sinuses.

The inferior alveolar nerve, a branch of the mandibular nerve, (a major division of the trigeminal nerve), enters the mandibular foramen and runs forward in the mandibular canal, supplying sensation to the teeth. At the mental foramen, the nerve divides into two terminal branches: incisive and mental nerves. The incisive nerve runs forward in the mandible and supplies the anterior teeth. The mental nerve exits the mental foramen and supplies sensation to the lower lip.

Variation

Males generally have squarer, stronger, and larger mandibles than females. The mental protuberance is more pronounced in males but can be visualized and palpated in females.

Rarely, a bifid inferior alveolar nerve may be present, in which case a second mandibular foramen, more inferiorly placed, exists and can be detected by noting a doubled mandibular canal on a radiograph.

Neck

The neck is the part of the body on many vertebrates that connects the head with the torso and provides the mobility and movements of the head. The structures of the human neck are anatomically grouped into four compartments; vertebral, visceral and two vascular compartments. Within these compartments, the neck houses the cervical vertebrae and cervical part of the spinal cord, upper parts of the respiratory and digestive tracts, endocrine glands, nerves, arteries and veins. Muscles of the neck are described separately from the compartments. They bound the neck triangles.

Structure

Compartments

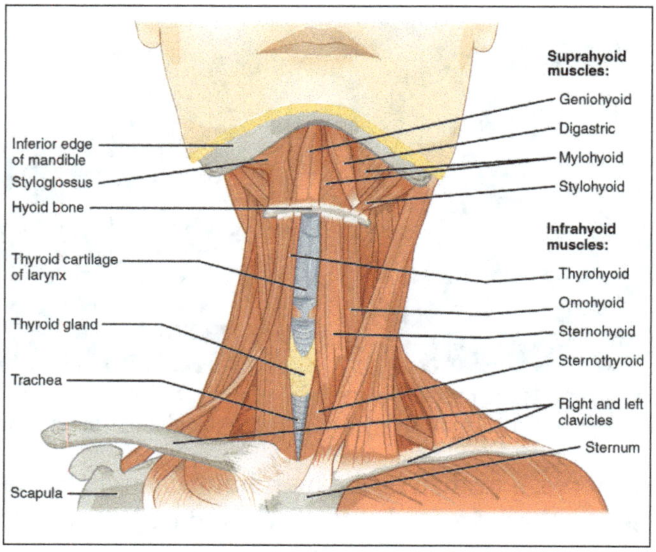

Muscles in the human neck.

The neck structures are distributed within four compartments:

- Vertebral compartment contains the cervical vertebrae with cartilaginous discs between each vertebral body. The alignment of the vertebrae defines the shape of the human neck. As the vertebrae bound the spinal canal, the cervical portion of the spinal cord is also found within the neck.
- Visceral compartment accommodates the trachea, larynx, pharynx, thyroid and parathyroid glands.
- Vascular compartment is paired and consists of the two carotid sheaths found on each side of the trachea. Each carotid sheath contains the vagus nerve, common carotid artery and internal jugular vein.

Besides the listed structures, the neck contains cervical lymph nodes which surround the blood vessels.

Muscles and Triangles

Muscles of the neck attach to the skull, hyoid bone, clavicles and the sternum. They bound the two major neck triangles; anterior and posterior.

- Anterior triangle is defined by the anterior border of the sternocleidomastoid muscle, inferior edge of the mandible and the midline of the neck. It contains the stylohyoid, digastric, mylohyoid, geniohyoid, omohyoid, sternohyoid, thyrohyoid and sternothyroid muscles. These muscles are grouped as the suprahyoid and infrahyoid muscles depending if they are located superiorly or inferiorly to the hyoid bone. The suprahyoid muscles (stylohyoid, digastric, mylohyoid, geniohyoid) elevate the hyoid bone, while the infrahyoid muscles (omohyoid, sternohyoid, thyrohyoid, sternothyroid) depress it. Acting synchronously, both groups facilitate speech and swallowing.

- Posterior triangle is bordered by the posterior border of the sternocleidomastoid muscle, anterior border of the trapezius muscle and the superior edge of the middle third of the clavicle. This triangle contains the sternocleidomastoid, trapezius, splenius capitis, levator scapulae, omohyoid, anterior, middle and posterior scalene muscles.

Nerve Supply

Sensation to the front areas of the neck comes from the roots of the spinal nerves C2-C4, and at the back of the neck from the roots of C4-C5. In addition to nerves coming from and within the human spine, the accessory nerve and vagus nerve travel down the neck.

Blood Supply and Vessels

Arteries which supply the neck are common carotid arteries which bifurcate into: Internal carotid artery - External carotid artery.

Surface Anatomy

The thyroid cartilage of the larynx forms a bulge in the midline of the neck called the Adam's apple. The Adam's apple is usually more prominent in men. Inferior to the Adam's apple is the cricoid cartilage. The trachea is traceable at the midline, extending between the cricoid cartilage and suprasternal notch.

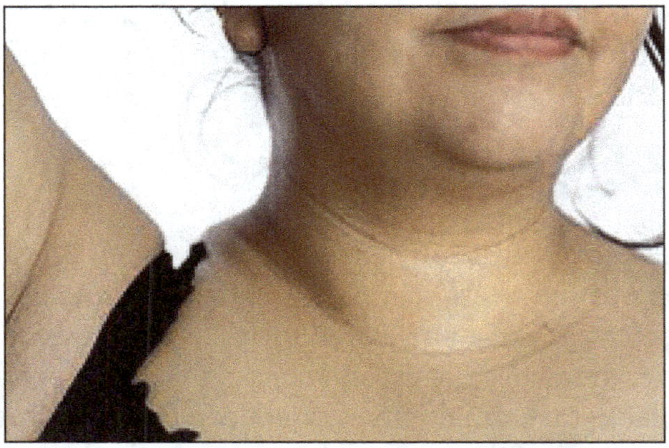

Development of neck lines (lat.monillas) or "moon rings" at mature age.

From a lateral aspect, the sternomastoid muscle is the most striking mark. It separates the anterior triangle of the neck from the posterior. The upper part of the anterior triangle contains the submandibular glands, which lie just below the posterior half of the mandible. The line of the common and the external carotid arteries can be marked by joining the sterno-clavicular articulation to the angle of the jaw. Neck lines appear at a later age as a development of skin wrinkles.

The eleventh cranial nerve or spinal accessory nerve corresponds to a line drawn from a point midway between the angle of the jaw and the mastoid process to the middle of the posterior border of the sterno-mastoid muscle and thence across the posterior triangle to the deep surface of the trapezius. The external jugular vein can usually be seen through the skin; it runs in a line drawn from the angle of the jaw to the middle of the clavicle, and close to it are some small lymphatic glands. The anterior jugular vein is smaller, and runs down about half an inch from the middle line of the neck. The clavicle or collar-bone forms the lower limit of the neck, and laterally the outward slope of the neck to the shoulder is caused by the trapezius muscle.

Common Carotid Arteries

The neck makes the connection between the head and the torso. The neck is made up of attachments from many muscles. The neck region contains many vital structures and organs that help regulate and maintain homeostasis of the body but most importantly, the impulses and regulation from the main control center of the body.

The neck is a very complex region of the body, and one of the most critical structures in the neck are the common carotids. The common carotids serve as one of the major blood supply to the head that originates from direct projection from the aorta. Along with the common carotid giving rise to blood flow to the head and brain from its branches, the common carotid also contains vital sensory organs that help control/regulate the blood pressure through the number of impulses using the autonomic nervous system.

The protection of the structures in the anterior neck, such as the airway, esophagus, blood vessels, and nerves are from the anterior neck muscles. These neck muscles protect these vital structures, but they also help support the head and influence facial expression and mastication.

Structure and Function

The overlying muscles protect the structures deep to them. The trachea is a midline structure that is deep to the overlying musculature. The trachea allows air to come and go into the lungs for gas exchange. Laterally to the midline and deep to the sternocleidomastoid muscle, there lays the carotid sheath that contains the common carotid artery, internal jugular vein, cranial nerves, and deep cervical lymph nodes. The common carotid arteries are one of the major pairs of arteries in the human body. The right common carotid artery is a branch of the brachiocephalic trunk while the left common carotid artery comes off the aorta directly. The common carotid arteries are considered large size muscular arteries. The artery is made up of three main layers. The most medial layer is called the tunica intima, and it is made up of epithelium and connective tissue. The layer superficial to the intima is the tunica media, and this layer is made up of smooth muscles. The smooth muscles provide the rigidity and facilitate pulse contraction to propel blood flow forward. The most superficial layer of the common carotids is the tunica adventitia, and this layer is made up of collagen and elastic tissue. This layer provides the flexibility and elasticity characteristics of the carotids. The characteristics of the pulsatile nature of the common carotid arteries also help propel lymph fluid in the nearby cervical lymph nodes that run close to the carotid sheath, draining the organs and structures of the head and neck.

The function of the common carotid arteries is to bring oxygenated blood from the left side of the heart to the head and neck. The common carotids travel in the carotid sheath and then bifurcate into the internal carotid and external carotid arteries at the level of cervical vertebrae four (C4). At the site of bifurcation, there are sensory organs called the carotid bodies. These sensory organs play a significant role in controlling blood pressure, oxygen level, and blood pH. The receptors specific for the blood pressure are the baroreceptors; these receptors sense the stretch and increase in pressure exerted on the carotid bodies. Baroreceptors send impulses to the brain through cranial nerve IX, and the brain sends an impulse back down through cranial nerve X. The baroceptors' impulses help influence the sympathetic nervous system to help maintain and regulate the blood pressure. The baroreceptors increase impulses when there is an increase in pressure or stretch in the carotid bodies, but when there is a decrease in stretch or pressure, the baroceptors decrease impulses to the central nervous system. The balance of blood control is carefully manipulated by the number of impulses from the baroreceptors, while the peripheral chemoreceptors respond to the increase in carbon dioxide or decrease the partial pressure of oxygen. When either of these two dissolved gases become disrupted, the peripheral chemoreceptors increase the respiratory rate to try and return the body to homeostasis.

Embryology

During the development of a fetus, the head and neck region partially develops from the brachial apparatus. The brachial apparatus is broken down into branchial clefts, pouches, and arches. The clefts are made up of ectoderm, but only the first cleft will form the external acoustic meatus while the rest of the clefts will obliterate. The brachial arch is composed of mesoderm and neural crest. The arches will eventually divide into the arteries, muscles, cartilage, and bony structures in the face and neck. The brachial pouches are made up of endoderm that will develop into the organs in the head and neck. Along with the development of the structures in the head and neck, there are aortic arches. The aortic arches will eventually develop into arteries for the head, face, neck, and branches from the aorta. The first aortic arch is responsible for developing the maxillary artery, which is one of the terminal branches from the external carotid artery. The second branch is the stapedial artery, which comes off the posterior auricular artery in the adult. The third aortic arch will form the common carotid artery, which will eventually fuse with the superior branches and become the main artery that supplies blood to the superior arteries derived from first and second aortic arches.

Blood Supply and Lymphatics

The blood supply to the common carotid arteries and main arteries is from the vasa vasorum. This a small network of the capillaries that supply the lumen and the walls of the common carotid arteries.

The common carotid arteries itself do not supply blood to the body, but its branches supply the head and neck regions from the external and internal carotid arteries. Following the bifurcation of the common carotid arteries, the internal carotid arteries go into the brain and supply the brain with blood via its terminal branches and contributing to the formation of the circle of Willis. The internal carotid artery branches into the anterior cerebral artery, and this supplies the frontal lobe medially and superiorly. The middle cerebral branch of the internal carotid supplies the lateral portion of the frontal lobes and the superior half of the temporal lobes. The blood supplies of the middle cerebral arteries are so vast that it even supplies some of the deep nuclei and the anterior portion of the parietal lobes.

Some of the smaller branches of the internal carotid like the posterior communicating arteries are mainly for collateral and bypass blood flow for the circle of Willis. The internal carotid artery also gives a branch off to supply the eyes as the ophthalmic arteries.

The external carotid arteries have eight major divisions that supply the head, neck, and face. As the external carotid arteries ascend the neck, it gives off a branch called the superior thyroid artery. The superior thyroid artery is vital for the muscles, organs, and structures predominantly inferior to the hyoid bone like the thyroid gland, infrahyoid muscles, and sternocleidomastoid muscle, but also the larynx above the hyoid bone. The second branch from the external carotid arteries is the ascending pharyngeal arteries. These arteries supply the middle ear, pharynx part of the oral cavity, prevertebral muscles and ascend to supply the meninges. The next branch that arises is the lingual artery that supplies the floor of the mouth and tongue. As the external carotid artery approach the mandible, it gives off the facial artery. The facial artery has multiple regions of the face, palate part of the oral cavity and tonsils, and submandibular glands that it supplies. The external carotid supplies the scalp through the next branch, which is the occipital artery. The posterior auricular artery is the last branch before the external carotid artery terminates into end branches. This artery is important in supplying the ear, scalp, cranial never VII, and parts of the scalp.

The external carotid artery terminates into the maxillary artery and superficial temporal artery. The maxillary supplies many structures in the head such as the muscles of mastication, teeth and the underlying gingivae, the jaw, the dura mater, the calvaria, the tympanic membrane, and external acoustic meatus of the ear. While the superficial temporal artery supplies the scalp around the temporal region.

Nerves

The common carotid artery's baroreceptors influence the regulation of the blood pressure. The baroreceptors send impulses to the brain via the Herring nerve, which is a division from cranial nerve IX (glossopharyngeal nerve). Once the impulses reach the central nervous system, the brain will respond by sending impulses to the sympathetic nervous system to increase or decrease in impulses via the sympathetic chain to either oppose the parasympathetic or increase parasympathetic impulses. The impulses from the parasympathetic nervous system are via cranial nerve X (vagus nerve) to the sinoatrial node and atrioventricular node to regulate blood pressure. The amount of impulse from the baroreceptors to the brain directly correlates to the increase of pressure or stretch on the common carotid arteries, but when the baroreceptors increase firing this leads to a decrease in efferent impulses from the sympathetic nervous system. When the sympathetic nervous system decreases in firing, unopposed parasympathetic impulses result and the opposite mechanism occurs when there is less pressure or stretch than normally exerted on the common carotid arteries.

While the common carotid arteries use the cranial nerves to control blood pressure, the nerves innervation for the muscles in the neck is motor or sensory. The platysma muscle is innervated by the cranial nerve VII (facial nerve) and assist in facial expression. The sternocleidomastoid muscle receives innervated from cranial nerve XI (spinal accessory nerve) which is important in head-turning. As for the nerve innervation of the suprahyoid muscles are different even though when these muscle work in sync to elevate the hyoid bone. The digastric muscle is unique in that aspect that it has two muscle bellies, and each muscle belly has a different nerve innervation. The anterior belly of the digastric muscle receives innervation by a branch from cranial nerve V (trigeminal nerve), but the posterior belly gets innervated by a branch of cranial nerve VII. The trigeminal nerve innervates the mylohyoid muscle along with the anterior belly of the digastric muscle. Along with innervating the posterior belly of the digastric muscle, the facial nerve also innervates the stylohyoid. As for the geniohyoid muscle, this is the only suprahyoid muscle that is innervated by cranial nerve XII (hypoglossal nerve).

The infrahyoid muscles are all innervated by the ansa cervicalis (C1-C3 nerve roots) except for the thyrohyoid muscle. The hypoglossal nerve innervates the thyrohyoid. All these muscles work in sync to depress the hyoid bone.

Muscles

The anterior structure of the neck is composed mainly of the complex network of neck muscles that protects the structures such as the trachea and common carotid arteries. The most superficial muscle is the platysma. The platysma muscles originate from the fascia above and below the neck. The platysma muscle blends in with the facial muscles. Due to the connection between muscles in the neck to muscles in the face, the platysma muscle helps with lowering the mandible and facial expression. Another superficial muscle in the neck is the sternocleidomastoid; this muscle functions to rotate the head to the contralateral side while side bending the head ipsilaterally. Along with the motion created by the sternocleidomastoid, it also protects the underlying carotid sheath and the contents within and nearby it. The sternocleidomastoid also functions to connect the head, neck, and thoracic through the attachments from the mastoid process to the sternum and clavicle.

Deep to the anterior superficial muscles, four paired muscles attach superior to the hyoid bone. These muscles are considered the suprahyoid muscles: digastric muscle, mylohyoid, geniohyoid, stylohyoid. The principal function of these muscles is to elevate the hyoid bone during speaking and swallowing. The digastric muscle has two bellies that are fused via the central tendon and originate from the mandible and attaches to the hyoid bone, allowing it to depress the mandible and elevate the hyoid at the same time during speaking and swallowing. The mylohyoid muscle extends from the inner surface of the mandible and projects to the central tendon to a fused structure called the midline raphe that attaches to the hyoid bone. The mylohyoid contributes to the depression of the mandible while elevating the hyoid bone and the floor of the mouth. The geniohyoid muscle starts from the inferior mental spines to the medial symphysis, and the stylohyoid starts from the styloid process to the hyoid bone. Both the geniohyoid muscle and the stylohyoid muscle will elevate the hyoid bone, but the geniohyoid muscle draws the hyoid bone anterior in motion while the stylohyoid pulls the bone posteriorly.

While inferior to the hyoid bone, there is another group of four paired muscles called the infrahyoid muscles: sternothyroid, sternohyoid, thyrohyoid, and omohyoid. The major function of these muscles depresses the hyoid bone and elevates the larynx to facilitate swallowing. While all the infrahyoid muscles depress the hyoid bone, the sternohyoid muscles also depress the larynx due to their shared origin from the anterior surface of the sternoclavicular joint to attach to the body of the hyoid bone. This is in contrast with the thyrohyoid muscle that elevates the larynx because it starts at the thyroid cartilage and attaches to the greater horn of the hyoid bone. The omohyoid muscle is unique because it has two bellies similar to the digastric muscle. This muscle starts from the superior border of the scapula and attaches to the body of the hyoid bone. Along with the attachment to the hyoid bone, the omohyoid muscle also serves a minor role in the protection of the carotid sheath's contents.

Physiologic Variants

The common carotid artery coming off the aorta on the left and the right common carotid artery usually comes off the brachiocephalic trunk. There are variations where the right common carotid artery comes directly from the aorta. Even rarer is the right common carotid artery originating from the right subclavian artery. The most common site for bifurcation of the common carotid artery is at C4, but the use of the C4 level as a landmark may be inconsistent in different individuals. The bifurcation may be slightly higher or lower than C4. The size of the internal carotid artery and external carotid artery can vary from person to person, but usually, the internal carotid artery tends to be slightly bigger. External carotid artery branches have many variations. The order of the eight branches coming off of the external carotid artery can vary greatly, but the region of blood supply stays consistent.

Adam's Apple

The Adam's apple is cartilage in the throat.

- The Adam's apple is a lump of cartilage that sticks out from the throat. It is more prominent in males, and it serves no specific function. A person can remove or change the size of their Adam's apple with surgery.
- Adam's apples develop during puberty and are not present in prepubescent children.
- They are often more prominent in males, who tend to develop larger larynxes than females. As the larynx grows behind the Adam's apple, it pushes the lump of cartilage outward.
- The size of the Adam's apple varies. Here, we describe what an Adam's apple is, why it develops, and options for removal or reduction.
- The Adam's apple, also known as the laryngeal prominence, is made of cartilage. This is a connective tissue present throughout the body.

- Cartilage is strong, but softer and more flexible than bone, and it makes up several other areas of the neck.
- The Adam's apple develops during puberty. It grows around the front of the larynx, just above the thyroid. It is usually more prominent in males.
- During puberty, the larynx grows and pushes the developing Adam's apple outward.
- The larynx, or voice box, is the area at the front of the neck that contains the vocal cords. These are folds of tissue that vibrate to produce sounds.
- Along with other body parts, such as the mouth and nasal passages, the larynx produces the voice. It also protects the airways during swallowing.
- The larynx grows extensively during puberty. A larger larynx produces deeper sounds, as there is more room for the vibrations to resonate.
- The cartilage around the larynx forms a shield to protect it from injury. The Adam's apple protrudes as the larynx grows, but it serves no individual function.
- Both males and females develop Adam's apples, but they tend to stick out more in males than in females.
- This is because the male larynx generally grows bigger and faster during puberty.
- A larger larynx explains why males tend to have deeper and louder voices than females.
- It is important to remember, however, that this is not always the case.
- Hormonal differences affect the way that the larynx, and Adam's apple, develop. As a result of these differences, a male may have a smaller larynx and a voice with a higher pitch.
- For the same reason, a female may have a larger larynx, a more prominent Adam's apple, and a voice with a lower pitch.

References

- Standring, Susan (2016). Gray's anatomy: the anatomical basis of clinical practice. Standring, Susan (Forty-first ed.). [Philadelphia]. ISBN 9780702052309. OCLC 920806541
- Head-anatomy, science: britannica.com, Retrieved 25 August, 2020
- Cassidy, Suzanne B.; Dykens, Elisabeth; Williams, Charles A. (2000). "Prader-Willi and Angelman syndromes: Sister imprinted disorders". American Journal of Medical Genetics. 97 (2): 136–46. doi:10.1002/1096-8628(200022)97:2<136::AID-AJMG5>3.0.CO;2-V. PMID 11180221
- Anatomy-eye, conditions-treatments: umkelloggeye.org, Retrieved 18 May, 2020
- Mitali Ruths (June 14, 2011). "White Line on Nose in Children". LiveStrong.com. Archived from the original on 2012-06-22. Retrieved 2012-11-25

4

Upper Limb Anatomy

The upper limb anatomy consists of three sections namely the upper arm, forearm, and hand that extends from the shoulder joint to the fingers. It also has many nerves, arteries, veins, and muscles. The topics elaborated in this chapter will help in gaining a better perspective of upper limb anatomy.

Arm

In zoology, arm is either of the forelimbs or upper limbs of ordinarily bipedal vertebrates, particularly humans and other primates. The term is sometimes restricted to the proximal part, from shoulder to elbow (the distal part is then called the forearm). In brachiating (tree-swinging) primates the arm is unusually long.

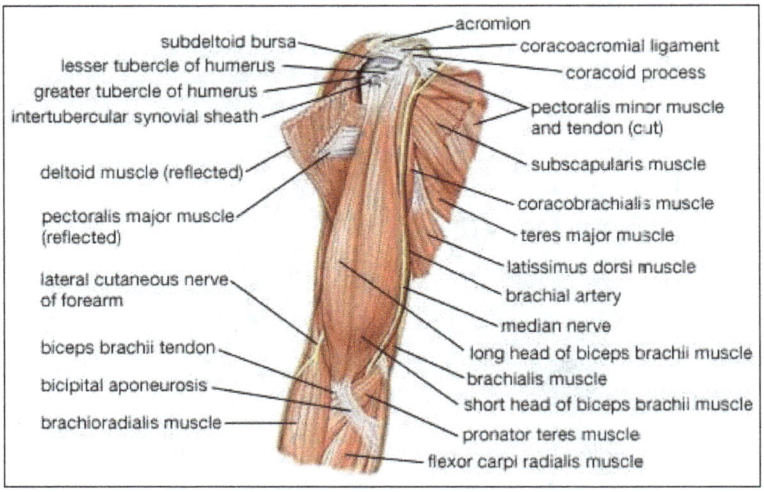

Muscles of the upper arm.

The bones of the human arm, like those of other primates, consist of one long bone, the humerus, in the arm proper; two thinner bones, the radius and ulna, in the forearm; and sets of carpal and metacarpal bones in the hand and digits in the fingers. The muscle that extends, or straightens, the arm is the triceps, which arises on the humerus and attaches to the ulna at the elbow; the brachialis and biceps muscles act to bend the arm at the elbow. A number of smaller muscles cover the radius and ulna and act to move the hand and fingers in various ways. The pectoralis muscle, anchored in the chest, is important in the downward motion of the entire arm and in quadrupeds pulls the limb backward in locomotion.

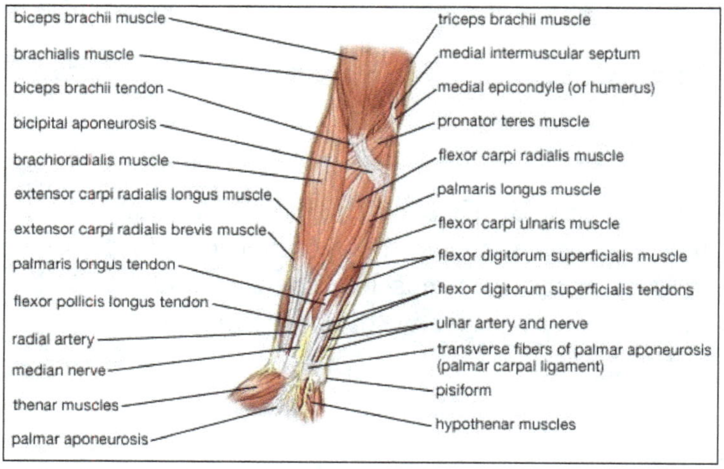

Muscles of the human forearm (anterior view, superficial layer).

Anatomy

Bones

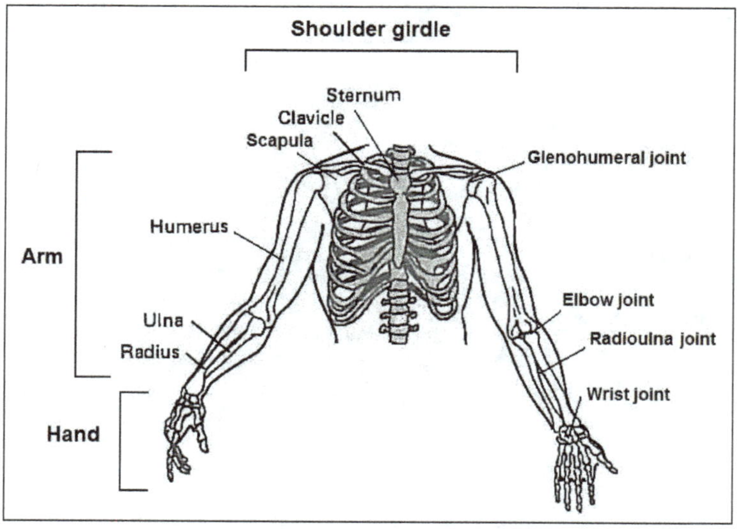

Bones of the upper limbs, together with shoulder girdles, that compose the human arm.

The humerus is one of the three long bones of the arm. It joins with the scapula at the shoulder joint and with the other long bones of the arm, the ulna and radius at the elbow joint. The elbow is a complex hinge joint between the end of the humerus and the ends of the radius and ulna.

Muscles

The arm is divided by a fascial layer (known as lateral and medial intermuscular septa) separating the muscles into two osteofascial compartments: the anterior and the posterior compartments of the arm. The fascia merges with the periosteum (outer bone layer) of the humerus.

The anterior compartment contains three muscles: biceps brachii, brachioradialis and coracobrachialis muscles. They are all innervated by the musculocutaneous nerve. The posterior compartment contains only the triceps brachii muscle, supplied by the radial nerve.

Nerve Supply

The musculocutaneous nerve, from C5, C6, C7, is the main supplier of muscles of the anterior compartment. It originates from the lateral cord of the brachial plexus of nerves. It pierces the coracobrachialis muscle and gives off branches to the muscle, as well as to brachialis and biceps brachii. It terminates as the anterior cutaneous nerve of the forearm.

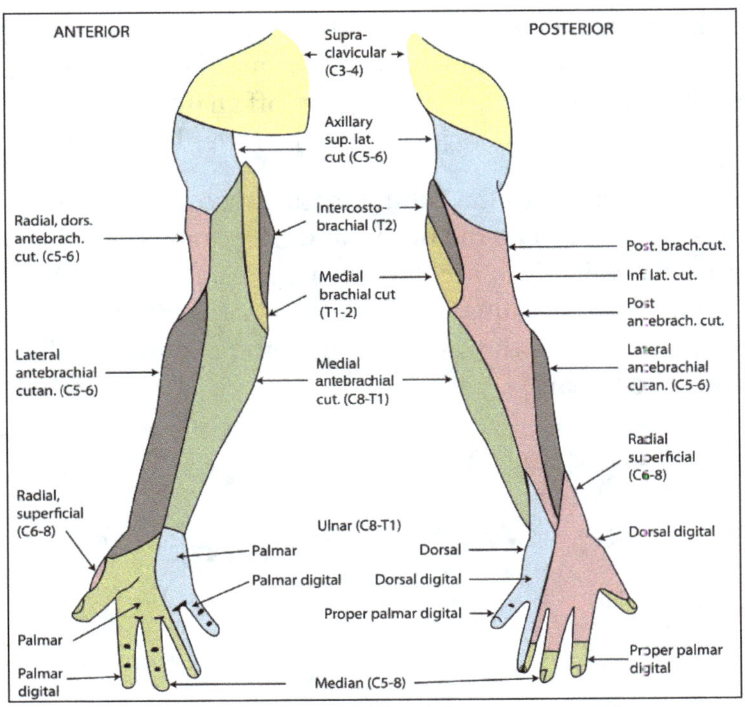

Cutaneous innervation of the right upper extremity.

The radial nerve, which is from the fifth cervical spinal nerve to the first thoracic spinal nerve, originates as the continuation of the posterior cord of the brachial plexus. This nerve enters the lower triangular space (an imaginary space bounded by, amongst others, the shaft of the humerus and the triceps brachii) of the arm and lies deep to the triceps brachii. Here it travels with the deep artery of the arm, which sits in the radial groove of the humerus. This fact is very important clinically as a fracture of the shaft of the bone here can cause lesions or even transections in the nerve.

Other nerves passing through give no supply to the arm. These include:

- The median nerve, nerve origin C5-T1, which is a branch of the lateral and medial cords of the brachial plexus. This nerve continues in the arm, travelling in a plane between the biceps and triceps muscles. At the cubital fossa, this nerve is deep to the pronator teres muscle and is the most medial structure in the fossa. The nerve passes into the forearm.
- The ulnar nerve, origin C8-T1, is a continuation of the medial cord of the brachial plexus. This nerve passes in the same plane as the median nerve, between the biceps and triceps muscles. At the elbow, this nerve travels posterior to the medial epicondyle of the humerus. This means that condylar fractures can cause lesion to this nerve.

Blood Supply

The main artery in the arm is the brachial artery. This artery is a continuation of the axillary artery. The point at which the axillary becomes the brachial is distal to the lower border of teres major. The brachial artery gives off an unimportant branch, the deep artery of arm. This branching occurs just below the lower border of teres major.

The brachial artery continues to the cubital fossa in the anterior compartment of the arm. It travels in a plane between the biceps and triceps muscles, the same as the median nerve and basilic vein. It is accompanied by venae comitantes (accompanying veins). It gives branches to the muscles of the anterior compartment. The artery is in between the median nerve and the tendon of the biceps muscle in the cubital fossa. It then continues into the forearm.

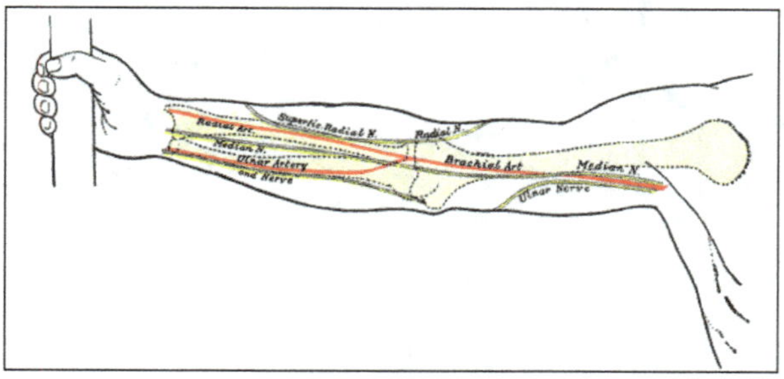

Main arteries of the arm.

The deep artery of the arm travels through the lower triangular space with the radial nerve. From here onwards it has an intimate relationship with the radial nerve. They are both found deep to the triceps muscle and are located on the spiral groove of the humerus. Therefore, fracture of the bone may not only lead to lesion of the radial nerve, but also haematoma of the internal structures of the arm. The artery then continues on to anastamose with the recurrent radial branch of the brachial artery, providing a diffuse blood supply for the elbow joint.

Veins

The veins of the arm carry blood from the extremities of the limb, as well as drain the arm itself. The two main veins are the basilic and the cephalic veins. There is a connecting vein between the two, the median cubital vein, which passes through the cubital fossa and is clinically important for venepuncture (withdrawing blood).

The basilic vein travels on the medial side of the arm and terminates at the level of the seventh rib.

The cephalic vein travels on the lateral side of the arm and terminates as the axillary vein. It passes through the deltopectoral triangle, a space between the deltoid and the pectoralis major muscles.

Elbow

The elbow is a hinge joint made up of the humerus, ulna and radius.

The unique positioning and interaction of the bones in the joint allows for a small amount of rotation as well as hinge action. This rotation is easily noticed during activities such as hand-to-mouth eating motions. The primary stability of the elbow is provided by the ulnar collateral ligament, on the medial (inner) side of the elbow. However, one of the most common injuries to the elbow occurs on the lateral, or outer, side of the elbow - it is called Lateral Epicondylitis, or Tennis Elbow.

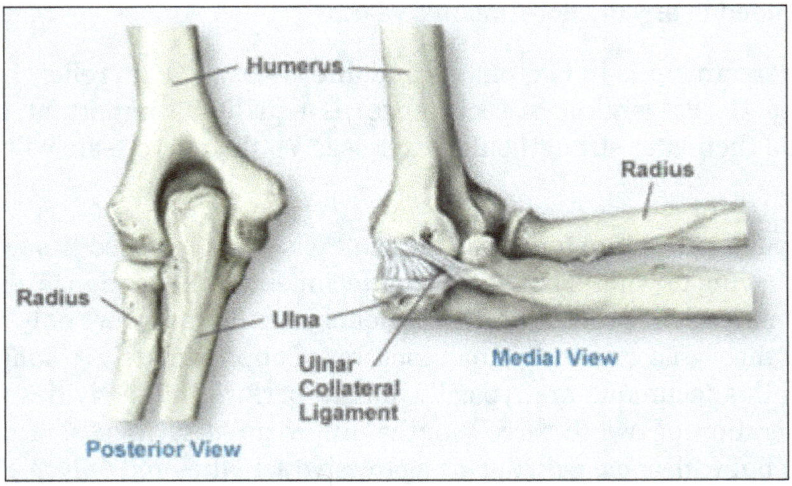

Anatomy of the Elbow.

Tennis Elbow (Lateral Epicondylitis)

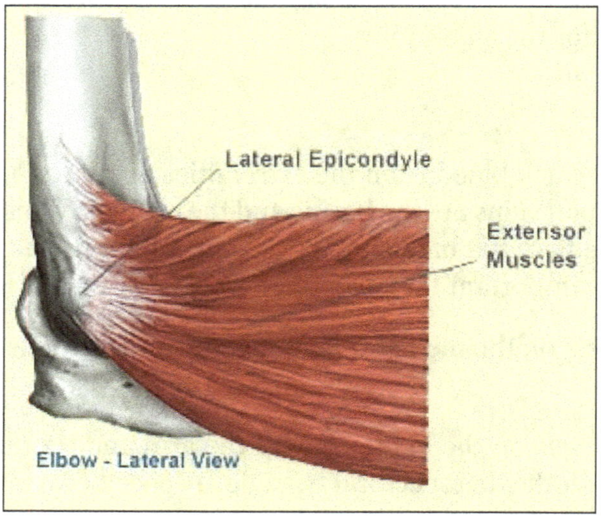

Tennis elbow, or lateral epicondylitis, is one of the most common elbow problems seen by an orthopedic surgeon. It is actually a tendinitis of the muscle called the extensor carpi radialis brevis which attaches to the lateral epicondyle of the humerus. It may be caused by a sudden injury or by repetitive use of the arm. Many doctors feel that micro tears in the tendon lead to a hyper-vascular phenomenon resulting in pain. The pain is usually worse with strong gripping with the elbow in an extended position, as in a tennis back hand stroke, but this problem can occur in golf and other sports as well as with repetitive use of tools.

Before surgery is considered a trial of at least six months of conservative treatment is indicated and may consist of a properly placed forearm brace and modification of elbow activities, anti-inflammatory medication and physical therapy. If the above treatment is not helpful, a cortisone injection can be beneficial but no more than three injections are recommended in any one location in a year.

Conservative treatment is in two phases and after Phase I (Pain relief) has been successful, Phase II (Prevention of recurrence) is equally as important and involves stretching and then later strengthening exercises, so the micro tears will not occur in the future.

The most popular procedure today is a simple excision of diseased tissue from within the tendon, shaving down the bone and re-attachment of the tendon. This can be performed as an outpatient procedure with regional anesthesia (where only the arm goes to sleep) and through a relatively small incision of approximately 3" long. 85-90% of patients with this technique are typically able to perform full activities without pain after a recuperation of two to three months. Approximately 10-12% of patients have improvement but with some pain during aggressive activities and only 2-3% of patients have no improvement.

Golfer's Elbow (Medial Epicondylitis)

Medial epicondylitis is inflammation of the tendon attachment of the flexor pronator muscles in the forearm. Usually this begins as microscopic tears in the tissue which leads to an inflammatory or hypervascular process. This occurs when stiff, underused tendons are suddenly overused or this may occur from an acute injury. The treatment includes three treatment options, no treatment, conservative and surgery.

Surgery is a last resort and involves cleaning up the tendon from diseased tissue, shaving down the bone and re-attachment of the tendon. This is necessary in 10-15% of the patients. Conservative treatment is in two phases, Phase I is to get rid of the pain and Phase II is to prevent it from coming back with stretching and strengthening exercises. To reduce the pain, using the elbow in a flexed position and the use of an elbow strap counterforce brace is usually the first line of treatment. If the patient has persistent symptoms a cortisone injection may be considered. No more than three injections are recommended per year and if the patient still has persistent symptoms despite conservative treatment, surgery is considered.

Cubital Tunnel Syndrome

Cubital tunnel syndrome is a pinched nerve at the elbow commonly known as the "funny bone". This might be caused by trauma or repetitive use of the elbow and may be caused by continuous use of the elbow in a flexed position. This causes the nerve to become stretched and irritated as opposed to when the arm is extended and the nerve is in a relaxed position. The diagnosis can be confirmed with electrodiagnostic testing including nerve conduction velocity and the electromyogram. Nerve conduction velocity studies, the speed of the nerve across the elbow, will be slowed when there is nerve compression and electromyogram studies, the innervation of the muscles, might be affected by the pinched nerve.

For this problem there are three modes of treatment; no treatment, conservative, and surgical. Unfortunately with conservative treatment, only splinting with the arm in an extended position has been found to be helpful. Night time splinting is achieved with a custom made long arm splint that the patient will wear at night time and as often as possible during the day. Unfortunately it is cumbersome to keep the arm out straight all the time and therefore this is usually used only at night.

If the patient has persistent complaints despite conservative treatment surgery would be recommended. There are three types of procedures, one is to cut the medial epicondyle which is the bone pinching the nerve or the other two operations are to actually move the nerve out of the cubital tunnel either above or below the muscles of the forearm. This can be performed as an outpatient procedure with an axillary block where only the arm is put to sleep and it has a high success rate.

Structure

Joint

The elbow joint has three different portions surrounded by a common joint capsule. These are joints between the three bones of the elbow, the humerus of the upper arm, and the radius and the ulna of the forearm.

Joint	From	To	Description
Humeroulnar joint	Trochlear notch of the ulna	Trochlea of humerus	is a simple hinge-joint, and allows for movements of flexion and extension only.
Humeroulnar joint	Head of the radius	Capitulum of the humerus	is a ball-and-socket joint.
Proximal radioulnar joint	Head of the radius	Radial notch of the ulna	in any position of flexion or extension, the radius, carrying the hand with it, can be rotated in it. This movement includes pronation and supination.

When in anatomical position there are four main bony landmarks of the elbow. At the lower part of the humerus are the medial and lateral epicondyles, on the side closest to the body (medial) and on the side away from the body (lateral) surfaces. The third landmark is the olecranon found at the head of the ulna. These lie on a horizontal line called the *Hueter line*. When the elbow is flexed, they form a triangle called the *Hueter triangle*, which resembles an equilateral triangle.

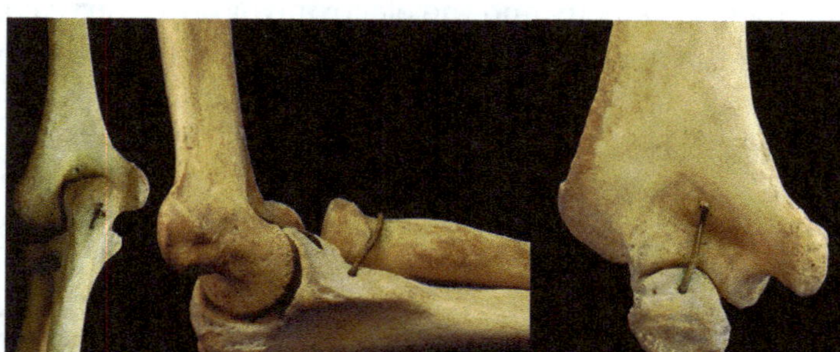

Left elbow extended and flexed.

At the surface of the humerus where it faces the joint is the trochlea. In most people, the groove running across the trochlea is vertical on the anterior side but it spirals off on the posterior side. These results in the forearm being aligned to the upper arm during flexion, but forming an angle to the upper arm during extension – an angle known as the carrying angle.

The superior radioulnar joint shares the joint capsule with the elbow joint but plays no functional role at the elbow.

Joint Capsule

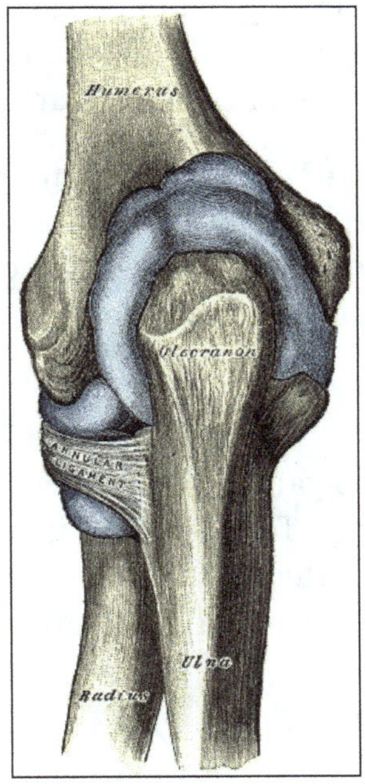

Capsule of elbow-joint (distended). Anterior and posterior aspects.

The elbow joint and the superior radioulnar joint are enclosed by a single fibrous capsule. The capsule is strengthened by ligaments at the sides but is relatively weak in front and behind.

On the anterior side, the capsule consists mainly of longitudinal fibres. However, some bundles among these fibers run obliquely or transversely, thickening and strengthening the capsule. These bundles are referred to as the capsular ligament. Deep fibres of the brachialis muscle insert anteriorly into the capsule and act to pull it and the underlying membrane during flexion in order to prevent them from being pinched.

On the posterior side, the capsule is thin and mainly composed of transverse fibres. A few of these fibres stretch across the olecranon fossa without attaching to it and form a transverse band with a free upper border. On the ulnar side, the capsule reaches down to the posterior part of the annular ligament. The posterior capsule is attached to the triceps tendon which prevents the capsule from being pinched during extension.

Synovial Membrane

The synovial membrane of the elbow joint is very extensive. On the humerus, it extends up from the articular margins and covers the coronoid and radial fossae anteriorly and the olecranon fossa posteriorly. Distally, it is prolonged down to the neck of the radius and the superior radioulnar joint. It is supported by the quadrate ligament below the annular ligament where it also forms a fold which gives the head of the radius freedom of movement.

Several synovial folds project into the recesses of the joint. These folds or plicae are remnants of normal embryonic development and can be categorized as either anterior (anterior humeral recess) or posterior (olecranon recess). A crescent-shaped fold is commonly present between the head of the radius and the capitulum of the humerus.

On the humerus there are extrasynovial fat pads adjacent to the three articular fossae. These pads fill the radial and coronoid fossa anteriorly during extension, and the olecranon fossa posteriorly during flexion. They are displaced when the fossae are occupied by the bony projections of the ulna and radius.

Ligaments

The elbow, like other joints, has ligaments on either side. These are triangular bands which blend with the joint capsule. They are positioned so that they always lie across the transverse joint axis and are, therefore, always relatively tense and impose strict limitations on abduction, adduction, and axial rotation at the elbow.

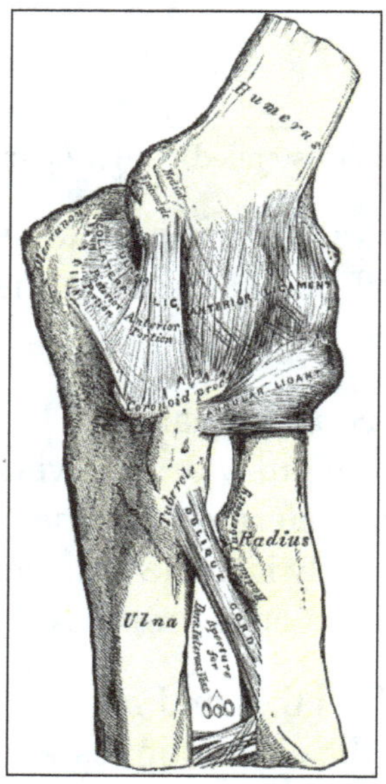

Left elbow-joint. Left: anterior and ulnar collateral ligaments.
Right: posterior and radial collateral ligaments.

The ulnar collateral ligament has its apex on the medial epicondyle. Its anterior band stretches from the anterior side of the medial epicondyle to the medial edge of the coronoid process, while the posterior band stretches from posterior side of the medial epicondyle to the medial side of the olecranon. These two bands are separated by a thinner intermediate part and their distal attachments are united by a transverse band below which the synovial membrane protrudes during joint movements. The anterior band is closely associated with the tendon of the superficial flexor muscles of the forearm, even being the origin of flexor digitorum superficialis. The ulnar nerve crosses the intermediate part as it enters the forearm.

The radial collateral ligament is attached to the lateral epicondyle below the common extensor tendon. Less distinct than the ulnar collateral ligament, this ligament blends with the annular ligament of the radius and its margins are attached near the radial notch of the ulna.

Muscles

Flexion

There are three main flexor muscles at the elbow:

- Brachialis acts exclusively as an elbow flexor and is one of the few muscles in the human body with a single function. It originates low on the anterior side of the humerus and is inserted into the tuberosity of the ulna.
- Brachioradialis acts essentially as an elbow flexor but also supinates during extreme pronation and pronates during extreme supination. It originates at the lateral supracondylar ridge distally on the humerus and is inserted distally on the radius at the styloid process.
- Biceps brachii is the main elbow flexor but, as a biarticular muscle, also plays important secondary roles as a stabiliser at the shoulder and as a supinator. It originates on the scapula with two tendons: That of the long head on the supraglenoid tubercle just above the shoulder joint and that of the short head on the coracoid process at the top of the scapula. Its main insertion is at the radial tuberosity on the radius.

Brachialis is the main muscle used when the elbow is flexed slowly. During rapid and forceful flexion all three muscles are brought into action assisted by the superficial forearm flexors originating at the medial side of the elbow. The efficiency of the flexor muscles increases dramatically as the elbow is brought into midflexion (flexed 90°) – biceps reaches its angle of maximum efficiency at 80–90° and brachialis at 100–110°.

Active flexion is limited to 145° by the contact between the anterior muscles of the upper arm and forearm, more so because they are hardened by contraction during flexion. Passive flexion (forearm is pushed against the upper arm with flexors relaxed) is limited to 160° by the bony projections on the radius and ulna as they reach to shallow depressions on the humerus; i.e. the head of radius being pressed against the radial fossa and the coronoid process being pressed against the coronoid fossa. Passive flexion is further limited by tension in the posterior capsular ligament and in triceps brachii.

A small accessory muscle, so called epitrochleoanconeus muscle, may be found on the medial aspect of the elbow running from the medial epicondyle to the olecranon.

Extension

Elbow extension is simply bringing the forearm back to anatomical position. This action is performed by triceps brachii with a negligible assistance from anconeus. Triceps originates with two heads posteriorly on the humerus and with its long head on the scapula just below the shoulder joint. It is inserted posteriorly on the olecranon.

Triceps is maximally efficient with the elbow flexed 20–30°. As the angle of flexion increases, the position of the olecranon approaches the main axis of the humerus which decreases muscle efficiency. In full flexion, however, the triceps tendon is "rolled up" on the olecranon as on a pulley which compensates for the loss of efficiency. Because triceps' long head is biarticular (acts on two joints), its efficiency is also dependent on the position of the shoulder.

Extension is limited by the olecranon reaching the olecranon fossa, tension in the anterior ligament, and resistance in flexor muscles. Forced extension results in a rupture in one of the limiting structures: Olecranon fracture, torn capsule and ligaments and though the muscles are normally left unaffected, a bruised brachial artery.

Blood Supply

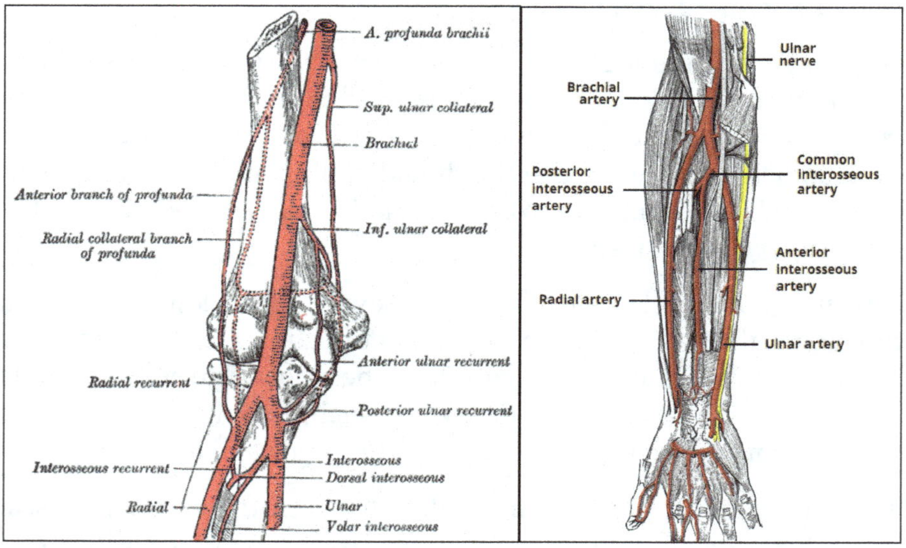

The anastomosis and deep veins around the elbow-joint.

The arteries supplying the joint are derived from an extensive circulatory anastomosis between the brachial artery and its terminal branches. The superior and inferior ulnar collateral branches of the brachial artery and the radial and middle collateral branches of the profunda brachii artery descend from above to reconnect on the joint capsule, where they also connect with the anterior and posterior ulnar recurrent branches of the ulnar artery; the radial recurrent branch of the radial artery; and the interosseous recurrent branch of the common interosseous artery.

The blood is brought back by vessels from the radial, ulnar, and brachial veins. There are two sets of lymphatic nodes at the elbow, normally located above the medial epicondyle — the deep and superficial cubital nodes (also called epitrochlear nodes). The lymphatic drainage at the elbow is through the deep nodes at the bifurcation of the brachial artery, the superficial nodes drain the forearm and the ulnar side of the hand. The efferent lymph vessels from the elbow proceed to the lateral group of axillary lymph nodes.

Nerve Supply

The elbow is innervated anteriorly by branches from the musculocutaneous, median, and radial nerve, and posteriorly from the ulnar nerve and the branch of the radial nerve to anconeus.

Development

The elbow undergoes dynamic development of ossification centers through infancy and adolescence, with the order of both the appearance and fusion of the apophyseal growth centers being crucial in assessment of the pediatric elbow on radiograph, in order to distinguish a traumatic fracture or apophyseal separation from normal development. The order of appearance can be understood by the mnemonic CRITOE, referring to the capitellum, radial head, internal epicondyle, trochlea, olecranon, and external epicondyle at ages 1, 3, 5, 7, 9 and 11 years. These apophyseal centers then fuse during adolescence, with the internal epicondyle and olecranon fusing last. The ages of fusion are more variable than ossification, but normally occur at 13, 15, 17, 13, 16 and 13 years, respectively. In addition, the presence of a joint effusion can be inferenced by the presence of the fat pad sign, a structure that is normally physiologically present, but pathologic when elevated by fluid, and always pathologic when posterior.

Function

The function of the elbow joint is to extend and flex the arm grasp and reach for objects. The range of movement in the elbow is from 0 degrees of elbow extension to 150 of elbow flexion. Muscles contributing to function are all flexion (biceps brachii, brachialis, and brachioradialis) and extension muscles (triceps and anconeus).

In humans, the main task of the elbow is to properly place the hand in space by shortening and lengthening the upper limb. While the superior radioulnar joint shares joint capsule with the elbow joint, it plays no functional role at the elbow.

With the elbow extended, the long axis of the humerus and that of the ulna coincide. At the same time, the articular surfaces on both bones are located in front of those axes and deviate from them at an angle of 45°. Additionally, the forearm muscles that originate at the elbow are grouped at the sides of the joint in order not to interfere with its movement. The wide angle of flexion at the elbow made possible by this arrangement — almost 180° — allows the bones to be brought almost in parallel to each other.

Carrying Angle

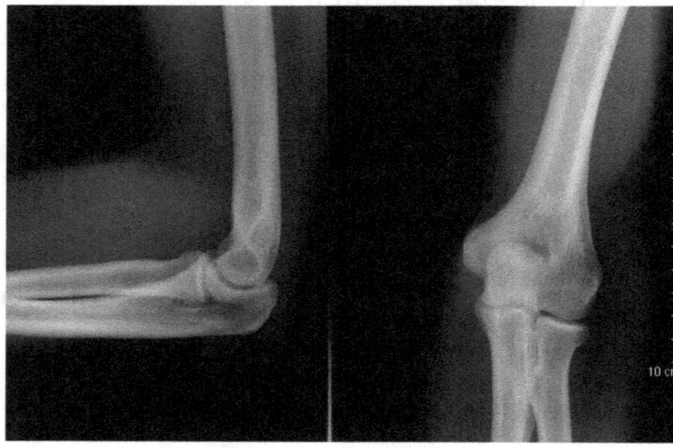

Normal radiograph; right picture of the straightened arm shows the carrying angle of the elbow.

When the arm is extended, with the palm facing forward or up, the bones of the upper arm (humerus) and forearm (radius and ulna) are not perfectly aligned. The deviation from a straight line occurs in the direction of the thumb, and is referred to as the "carrying angle" (visible in the right half of the picture, right).

The carrying angle permits the arm to be swung without contacting the hips. Women on average have smaller shoulders and wider hips than men, which tends to produce a larger carrying angle (i.e., larger deviation from a straight line than that in men). There is, however, extensive overlap in the carrying angle between individual men and women, and a sex-bias has not been consistently observed in scientific studies. This could however be attributed to the very small sample sizes in those cited earlier studies. A more recent study based on a sample size of 333 individuals from both sexes concluded that carrying angle is a suitable secondary sexual characteristic.

The angle is greater in the dominant limb than the non-dominant limb of both sexes, suggesting that natural forces acting on the elbow modify the carrying angle. Developmental, aging and possibly racial influences add further to the variability of this parameter.

Forearm

The forearm is the region of the upper limb between the elbow and the wrist. The term forearm is used in anatomy to distinguish it from the arm, a word which is most often used to describe the entire appendage of the upper limb, but which in anatomy, technically, means only the region of the upper arm, whereas the lower "arm" is called the forearm. It is homologous to the region of the leg that lies between the knee and the ankle joints, the crus.

The forearm contains two long bones, the radius and the ulna, forming the radioulnar joint. The interosseous membrane connects these bones. Ultimately, the forearm is covered by skin, the anterior surface usually being less hairy than the posterior surface.

The forearm contains many muscles, including the flexors and extensors of the digits, a flexor of the elbow (brachioradialis), and pronators and supinators that turn the hand to face down or upwards, respectively. In cross-section the forearm can be divided into two fascial compartments. The posterior compartment contains the extensors of the hands, which are supplied by the radial nerve. The anterior compartment contains the flexors, and is mainly supplied by the median nerve. The flexor muscles are more massive than the extensors, because they work against gravity and act as anti-gravity muscles. The ulnar nerve also runs the length of the forearm.

The radial and ulnar arteries and their branches supply the blood to the forearm. These usually run on the anterior face of the radius and ulna down the whole forearm. The main superficial veins of the forearm are the cephalic, median antebrachial and the basilic vein. These veins can be used for cannularisation or venipuncture, although the cubital fossa is a preferred site for getting blood.

Hand

A hand is a prehensile, multi-fingered appendage located at the end of the forearm or forelimb of primates such as humans, chimpanzees, monkeys, and lemurs. A few other vertebrates such as the koala (which has two opposable thumbs on each "hand" and fingerprints extremely similar to human fingerprints) are often described as having "hands" instead of paws on their front limbs. The raccoon is usually described as having "hands" though opposable thumbs are lacking.

Some evolutionary anatomists use the term hand to refer to the appendage of digits on the forelimb more generally — for example, in the context of whether the three digits of the bird hand involved the same homologous loss of two digits as in the dinosaur hand.

The human hand normally has five digits: four fingers plus one thumb; these are often referred to collectively as five fingers, however, whereby the thumb is included as one of the fingers. It has 27 bones, not including the sesamoid bone, the number of which varies between people, 14 of which are the phalanges (proximal, intermediate and distal) of the fingers and thumb. The metacarpal bones connect the fingers and the carpal bones of the wrist. Each human hand has five metacarpals and eight carpal bones.

Fingers contain some of the densest areas of nerve endings in the body, and are the richest source of tactile feedback. They also have the greatest positioning capability of the body; thus, the sense of touch is intimately associated with hands. Like other paired organs (eyes, feet, legs) each hand is dominantly controlled by the opposing brain hemisphere, so that handedness—the preferred hand choice for single-handed activities such as writing with a pencil, reflects individual brain functioning.

Among humans, the hands play an important function in body language and sign language. Likewise the ten digits of two hands, and the twelve phalanges of four fingers (touchable by the thumb) have given rise to number systems and calculation techniques.

Structure

Many mammals and other animals have grasping appendages similar in form to a hand such as paws, claws, and talons, but these are not scientifically considered to be grasping hands. The scientific use of the term *hand* in this sense to distinguish the terminations of the front paws from the hind ones is an example of anthropomorphism. The only true grasping hands appear in the mammalian order of primates. Hands must also have opposable thumbs.

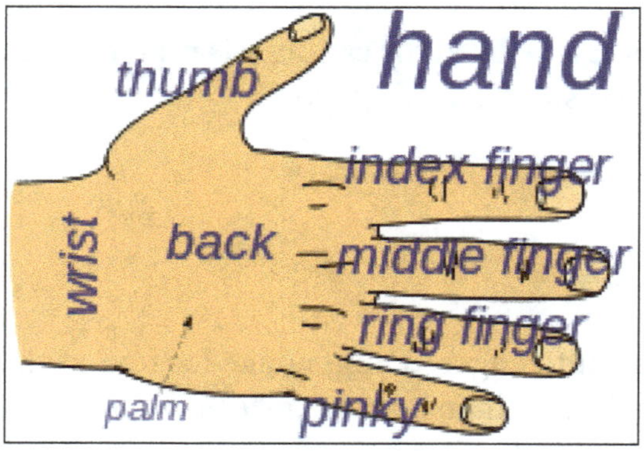

Human hand parts.

The hand is located at the distal end of each arm. Apes and monkeys are sometimes described as having four hands, because the toes are long and the hallux is opposable and looks more like a thumb, thus enabling the feet to be used as hands.

The word "hand" is sometimes used by evolutionary anatomists to refer to the appendage of digits on the forelimb such as when researching the homology between the three digits of the bird hand and the dinosaur hand. An adult human male's hand weighs about a pound.

Areas of the human hand include:

- The palm (Volar), which is the central region of the anterior part of the hand, located superficially to the metacarpus. The skin in this area contains dermal papillae to increase friction, such as are also present on the fingers and used for fingerprints.
- The opisthenar area (dorsal) is the corresponding area on the posterior part of the hand.

- The heel of the hand is the area anteriorly to the bases of the metacarpal bones, located in the proximal part of the palm. It is the area that sustains most pressure when using the palm of the hand for support, such as in handstand.

There are five digits attached to the hand, notably with a nail fixed to the end in place of the normal claw. The four fingers can be folded over the palm which allows the grasping of objects. Each finger, starting with the one closest to the thumb, has a colloquial name to distinguish it from the others:

- Index finger, pointer finger, forefinger, or 2nd digit.
- Middle finger or long finger or 3rd digit.
- Ring finger or 4th digit.
- Little finger, pinky finger, small finger, baby finger, or 5th digit.

The thumb (connected to the first metacarpal bone and trapezium) is located on one of the sides, parallel to the arm. A reliable way of identifying human hands is from the presence of opposable thumbs. Opposable thumbs are identified by the ability to be brought opposite to the fingers, a muscle action known as opposition.

Bones

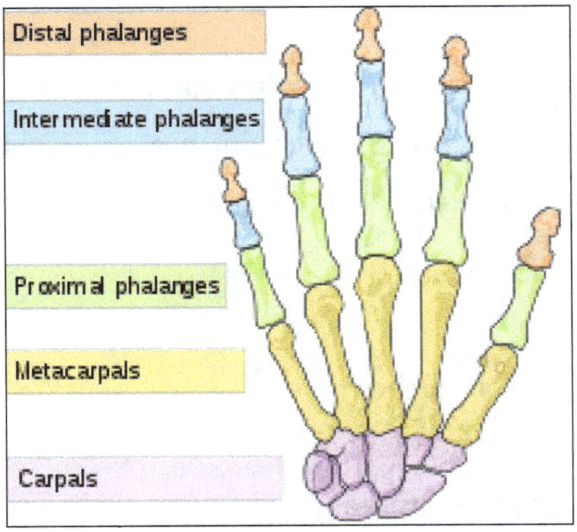

Bones of the human hand.

The skeleton of the human hand consists of 27 bones: the eight short carpal bones of the wrist are organized into a proximal row (scaphoid, lunate, triquetral and pisiform) which articulates with the bones of the forearm, and a distal row (trapezium, trapezoid, capitate and hamate), which articulates with the bases of the five metacarpal bones of the hand. The heads of the metacarpals will each in turn articulate with the bases of the proximal phalanx of the fingers and thumb. These articulations with the fingers are the metacarpophalangeal joints known as the knuckles. At the palmar aspect of the first metacarpophalangeal joints are small, almost spherical bones called the sesamoid bones. The fourteen phalanges make up the fingers and thumb, and are numbered I-V (thumb to little finger) when the hand is viewed from an anatomical position (palm up). The four fingers each consist of three phalanx bones: proximal, middle, and distal. The thumb only consists of a proximal and distal phalanx. Together with the phalanges of the fingers and thumb these metacarpal bones form five rays or poly-articulated chains.

Because supination and pronation (rotation about the axis of the forearm) are added to the two axes of movements of the wrist, the ulna and radius are sometimes considered part of the skeleton of the hand.

There are numerous sesamoid bones in the hand, small ossified nodes embedded in tendons; the exact number varies between people: whereas a pair of sesamoid bones are found at virtually all thumb metacarpophalangeal joints, sesamoid bones are also common at the interphalangeal joint of the thumb (72.9%) and at the metacarpophalangeal joints of the little finger (82.5%) and the index finger (48%). In rare cases, sesamoid bones have been found in all the metacarpophalangeal joints and all distal interphalangeal joints except that of the long finger.

The articulations are:

- Interphalangeal articulations of hand (the hinge joints between the bones of the digits).
- Metacarpophalangeal joints (where the digits meet the palm).
- Intercarpal articulations (where the palm meets the wrist).
- Wrist (may also be viewed as belonging to the forearm).

Arches

Arches of the hand:

- Red: One of the oblique arches.
- Brown: One of the longitudinal arches of the digits.
- Dark green: Transverse carpal arch.
- Light green: Transverse metacarpal arch.

The fixed and mobile parts of the hand adapt to various everyday tasks by forming bony arches: longitudinal arches (the rays formed by the finger bones and their associated metacarpal bones), transverse arches (formed by the carpal bones and distal ends of the metacarpal bones), and oblique arches (between the thumb and four fingers).

Of the longitudinal arches or rays of the hand, that of the thumb is the most mobile (and the least longitudinal). While the ray formed by the little finger and its associated metacarpal bone still offers some mobility, the remaining rays are firmly rigid. The phalangeal joints of the index finger, however, offer some independence to its finger, due to the arrangement of its flexor and extension tendons.

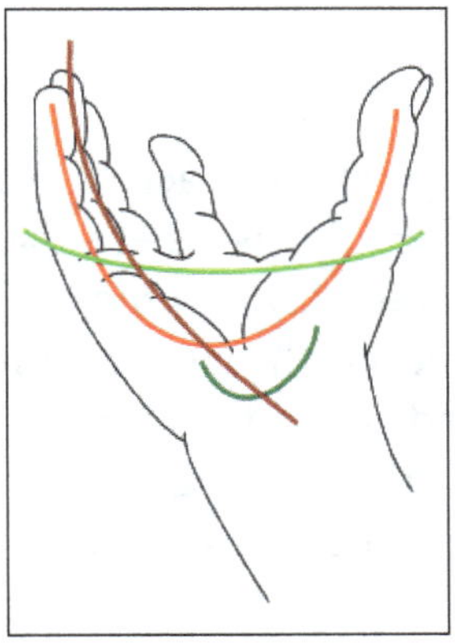

The carpal bones form two transversal rows, each forming an arch concave on the palmar side. Because the proximal arch simultaneously has to adapt to the articular surface of the radius and to the distal carpal row, it is by necessity flexible. In contrast, the capitate, the "keystone" of the distal arch, moves together with the metacarpal bones and the distal arch is therefore rigid. The stability of these arches is more dependent of the ligaments and capsules of the wrist than of the interlocking shapes of the carpal bones, and the wrist is therefore more stable in flexion than in extension. The distal carpal arch affects the function of the CMC joints and the hands, but not the function of the wrist or the proximal carpal arch. The ligaments that maintain the distal carpal arches are the transverse carpal ligament and the intercarpal ligaments (also oriented transversally). These ligaments also form the carpal tunnel and contribute to the deep and superficial palmar arches. Several muscle tendons attaching to the TCL and the distal carpals also contribute to maintaining the carpal arch.

Compared to the carpal arches, the arch formed by the distal ends of the metacarpal bones is flexible due to the mobility of the peripheral metacarpals (thumb and little finger). As these two metacarpals approach each other, the palmar gutter deepens. The central-most metacarpal (middle finger) is the most rigid. It and its two neighbors are tied to the carpus by the interlocking shapes of the metacarpal bones. The thumb metacarpal only articulates with the trapezium and is therefore completely independent, while the fifth metacarpal (little finger) is semi-independent with the fourth metacarpal (ring finger) which forms a transitional element to the fifth metacarpal.

Together with the thumb, the four fingers form four oblique arches, of which the arch of the index finger functionally is the most important, especially for precision grip, while the arch of the little finger contribute an important locking mechanism for power grip. The thumb is undoubtedly the "master digit" of the hand, giving value to all the other fingers. Together with the index and middle finger, it forms the dynamic tridactyl configuration responsible for most grips not requiring force. The ring and little fingers are more static, a reserve ready to interact with the palm when great force is needed.

Shoulder

The shoulder is made up of two joints, the acromioclavicular joint and the glenohumeral joint. The acromioclavicular joint is where the acromion, part of the shoulder blade (scapula) and the collar bone (clavicle) meet. The glenohumeral joint is where the ball (humeral head) and the socket (the glenoid) meet. The rotator cuff connects the humerus to the scapula and is made up of the tendons of four muscles, the supraspinatus, infraspinatus, teres minor and the subscapularis. Tendons attach muscle to bone. Muscles in turn move bones by pulling on the tendons. The muscles of the rotator cuff keep the humerus tightly in the socket. The socket, or the glenoid, is shallow and flat. It is rimmed with soft tissue called the labrum that makes a deeper socket that molds to fit the humeral head. The joint capsule surrounds the shoulder joint. It is a fluid filled sac that lubricates the joint. It is made up of ligaments. Ligaments are soft tissue that holds bone to bone. Shoulder injuries can occur to any part of the shoulder.

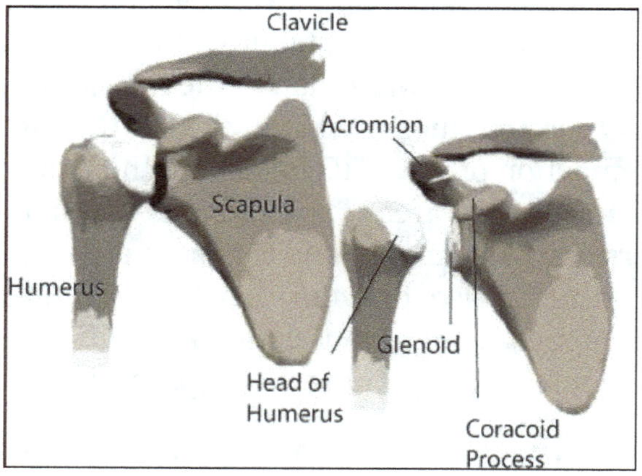

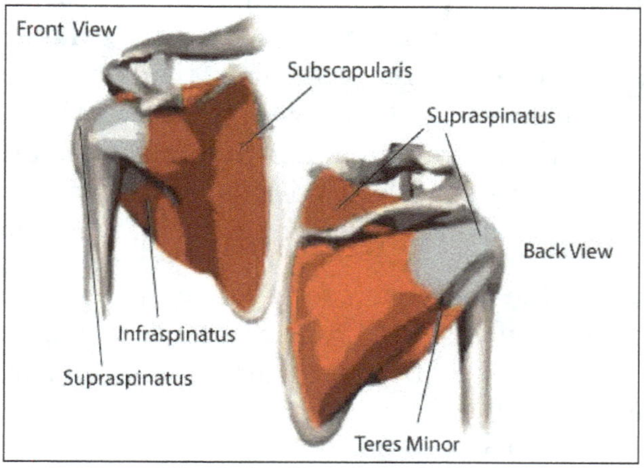

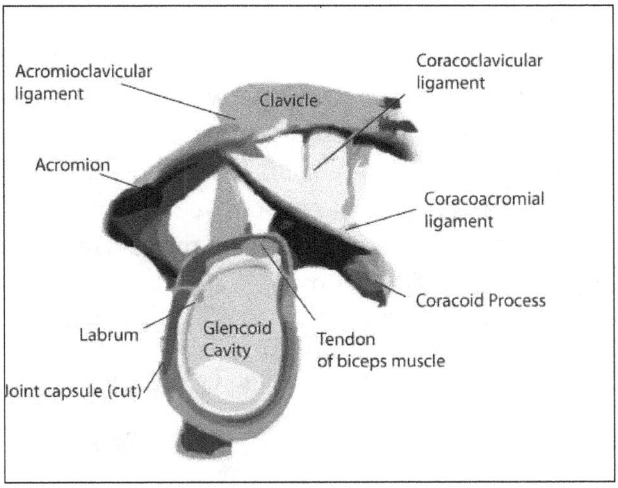

Structure

The shoulder consists of a ball-and-socket joint formed by the humerus and scapula and their surrounding structures - ligaments, muscles, tendons - which support the bones and maintain the relationship of one to another. These supporting structures attach to the clavicle, humerus, and scapula, the latter providing the glenoid cavity, acromion and coracoid processes. The main joint of the shoulder is the shoulder joint (or glenohumeral joint), between the humerus and the glenoid process of the scapular. The acromioclavicular joint and sternoclavicular joint also play a role in shoulder movements. White hyaline cartilage on the ends of the bones (called articular cartilage) allows the bones to glide and move on each other, and the joint space is surrounded by a synovial membrane. Around the joint space are muscles - the rotator cuff, which directly surrounds and attaches to the shoulder joint - and other muscles that help provide stability and facilitate movement.

Two filmy sac-like structures called bursae permit smooth gliding between bone, muscle, and tendon. They cushion and protect the rotator cuff from the bony arch of the acromion.

The glenoid labrum is the second kind of cartilage in the shoulder which is distinctly different from the articular cartilage. This cartilage is more fibrous or rigid than the cartilage on the ends of the ball and socket. Also, this cartilage is also found only around the socket where it is attached.

Shoulder Joint

The shoulder joint (also known as the glenohumeral joint) is the main joint of the shoulder. It is a ball and socket joint that allows the arm to rotate in a circular fashion or to hinge out and up away from the body. It is formed by the articulation between the head of the humerus and the lateral scapula (specifically-the glenoid cavity of the scapula). The "ball" of the joint is the rounded, medial anterior surface of the humerus and the "socket" is formed by the glenoid cavity, the dish-shaped portion of the lateral scapula. The shallowness of the cavity and relatively loose connections between the shoulder and the rest of the body allows the arm to have tremendous mobility, at the expense of being much easier to dislocate than most other joints in the body. There is an approximately 4-to-1 disproportion in size between the large head of the humerus and the shallow glenoid cavity. The glenoid cavity is made deeper by the addition of the fibrocartilaginous ring of the glenoid labrum.

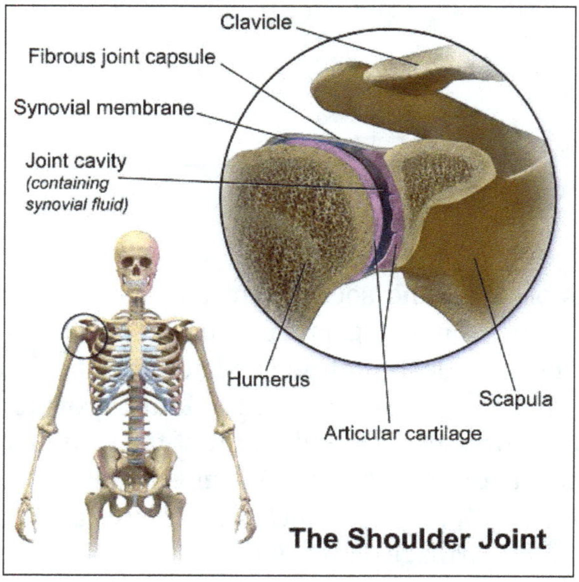

Cross-section of shoulder joint.

The capsule is a soft tissue envelope that encircles the glenohumeral joint and attaches to the scapula, humerus, and head of the biceps. It is lined by a thin, smooth synovial membrane. This capsule is strengthened by the coracohumeral ligament which attaches the coracoid process of the scapula to the greater tubercle of the humerus. There are also three other ligaments attaching the lesser tubercle of the humerus to lateral scapula and are collectively called the glenohumeral ligaments.

The transverse humeral ligament, which passes from the lesser tubercle to the greater tubercle of humerus, covers the intertubercular groove, in which the long head of biceps brachii travels.

Rotator Cuff

The rotator cuff is an anatomical term given to the group of four muscles and their tendons that act to stabilize the shoulder. These muscles are the supraspinatus, infraspinatus, teres minor and subscapularis and that hold the head of the humerus in the glenoid cavity during movement. The cuff adheres to the glenohumeral capsule and attaches to the head of the humerus. Together, these keep the humeral head in the glenoid cavity, preventing upward migration of the humeral head caused by the pull of the deltoid muscle at the beginning of arm elevation. The infraspinatus and the teres minor, along with the anterior fibers of the deltoid muscle, are responsible for external rotation of the arm.

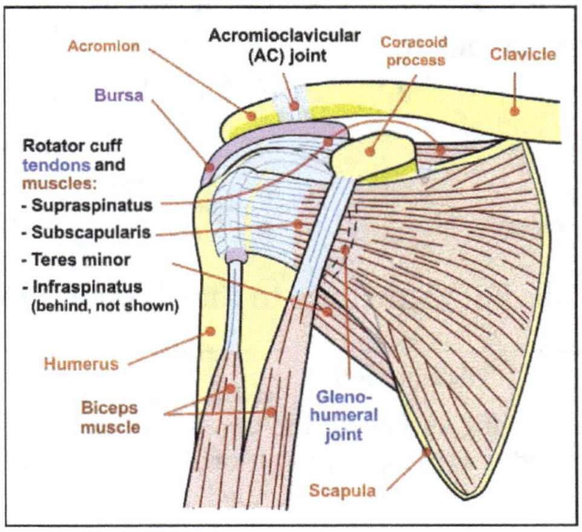

Shoulder anatomy, front view.

The four tendons of these muscles converge to form the rotator cuff tendon. This tendon, along with the articular capsule, the coracohumeral ligament, and the glenohumeral ligament complex, blend into a confluent sheet before insertion into the humeral tuberosities. The infraspinatus and teres minor fuse near their musculotendinous junctions, while the supraspinatus and subscapularis tendons join as a sheath that surrounds the biceps tendon at the entrance of the bicipital groove.

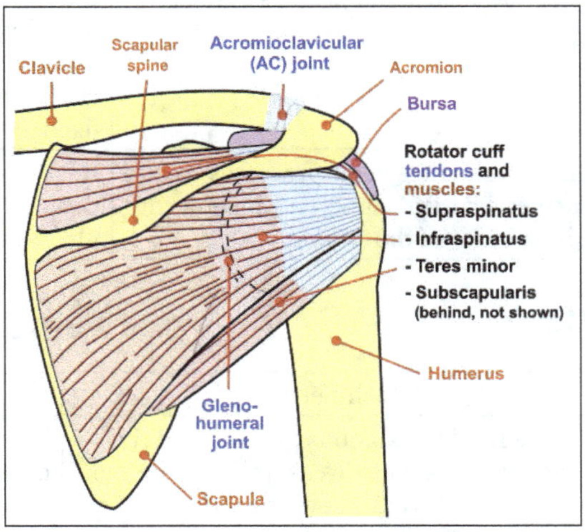

Shoulder anatomy, back view.

Other Muscles

Muscles from the Shoulder Region

In addition to the four muscles of the rotator cuff, the deltoid muscle and teres major muscles arise and exist in the shoulder region itself. The deltoid muscle covers the shoulder joint on three sides, arising from the front upper third of the clavicle, the acromion, and the spine of the scapula, and travelling to insert on the deltoid tubercle of the humerus. Contraction of each part of the deltoid assists in different movements of the shoulder - flexion (clavicular part), abduction (middle part) and extension (scapular part). The teres major attaches to the outer part of the back of the scapula, beneath the teres minor, and attaches to the upper part of the humerus. It helps with medial rotation of the humerus.

Muscles from the Front

Muscles from the chest wall that contribute to the shoulder are:

Name	Attachment	Function
Serratus anterior	Originates on the surface of the upper eight ribs at the side of the chest and inserts along the entire anterior length of the medial border of the scapula.	It fixes the scapula into the thoracic wall and aids in rotation and abduction of the shoulders.
Subclavius	Located beneath the clavicle, originating from the first rib and inserting on the subclavian groove of the clavicle.	It depresses the lateral clavicle and also acts to stabilize the clavicle.
Pectoralis minor	Arises from the third, fourth, and fifth ribs, near their cartilage and inserts into the medial border and upper surface of the coracoid process of the scapula.	This muscle aids in respiration, medially rotates the scapula, protracts the scapula, and also draws the scapula inferiorly.
Sternocleidomastoid	Attaches to the sternum (sterno-), the clavicle (cleido-), and the mastoid process of the temporal bone of the skull.	Most of its actions flex and rotate the head. In regards to the shoulder, however, it also aids in respiration by elevating the sternoclavicular joint when the head is fixed.
Levator scapulae	Arises from the transverse processes of the first four cervical vertebrae and inserts into the medial border of the scapula.	It is capable of rotating the scapula downward and elevating the scapula.

Muscles from the Back

Name	Attachment	Function
Rhomboid major and rhomboid minor (work together)	They arise from the spinous processes of the thoracic vertebrae T1 to T5 as well as from the spinous processes of the seventh cervical. They attach to the inner border of the scapula.	They are responsible for downward rotation of the scapula with the levator scapulae, as well as adduction of the scapula.
Trapezius	Arises from the occipital bone, the ligamentum nuchae, the spinous process of the seventh cervical, and the spinous processes of all the thoracic vertebrae. It attaches to the outer clavicle, the acromion process, and into the spine of the scapula.	Different portions of the fibers perform different actions on the scapula: depression, upward rotation, elevation, and retraction.

Levator scapulae	Arises from the transverse processes of cervical vertebrae 1-4, and attaches to the upper part of the inner border of the scapula.	Elevates the scapula.
Latissimus dorsi	A large muscle that arises form the spinous processes of the lower six thoracic vertebrae, lumbar and all sacral vertebrae, and posterior iliac crest. It attaches to the intertubercular groove of the humerus.	Adducts, extends and rotates the humerus inwards.

Armpit

The armpit is formed by the space between the muscles of the shoulder. The nerves and blood vessels of the arm travel through the armpit, and it possesses several sets of lymph nodes that are able to be examined. The armpit is formed by the pectoralis major and minor muscles at the front, the latissimus dorsi and teres major muscles at the back, the serratus anterior muscle on its inner surface, and the intertubercular groove of the humerus on the outer side.

Nerve Supply and Passage

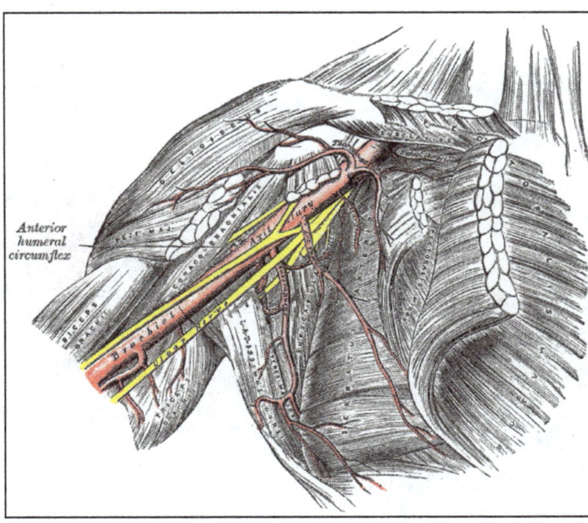

The brachial plexus surrounds the axillary artery and is formed from nerve roots from C5-T1. Branches of the plexus supply the shoulder region, although several other nerves play a role.

The skin around the shoulder is supplied by C2-C4 (upper), and C7 and T2 (lower area). The brachial plexus emerges as nerve roots from the cervical vertebrae C5-T1. Branches of the plexus, in particular from C5-C6, supply the majority of the muscles of the shoulder.

Blood Vessels

The subclavian artery arises from the brachiocephalic trunk on the right and directly from the aorta from the left. This becomes the axillary artery as it passes beyond the first rib.

The axillary artery also supplies blood to the arm, and is one of the major sources of blood to the shoulder region. The other major sources are the transverse cervical artery and the suprascapular artery, both branches of the thyrocervical trunk which itself is a branch of the subclavian artery. The blood vessels form a network (anastamosis) behind the shoulder that helps to supply blood to the arm even when the axillary artery is compromised.

Muscles of the Upper Limb

Trapezius

The trapezius is a large paired surface muscle that extends longitudinally from the occipital bone to the lower thoracic vertebrae of the spine and laterally to the spine of the scapula. It moves the scapula and supports the arm.

The trapezius has three functional parts: an upper (descending) part which supports the weight of the arm; a middle region (transverse), which retracts the scapula; and a lower (ascending) part which medially rotates and depresses the scapula.

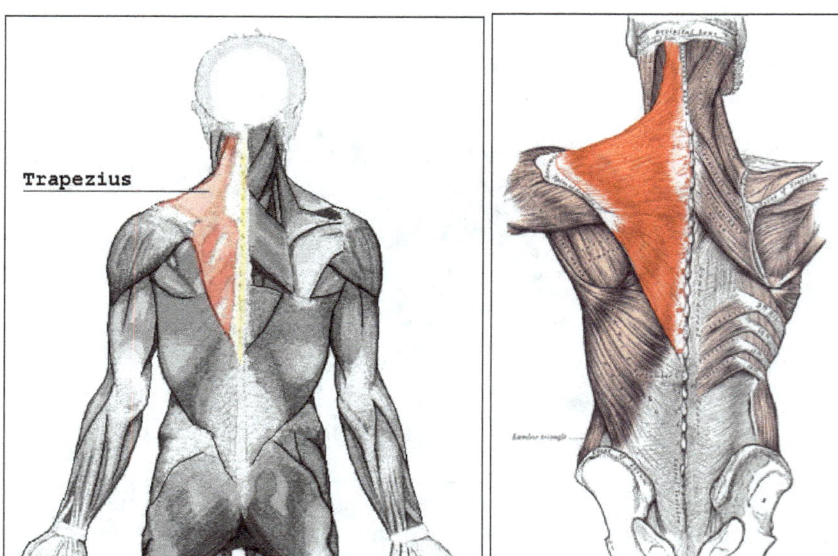

The trapezius muscle is a surface muscle of back, shown in red.

Structure

The trapezius muscle resembles a trapezium (trapezoid in American English), or diamond-shaped quadrilateral. The word "spinotrapezius" refers to the human trapezius, although it is not commonly used in modern texts. In other mammals, it refers to a portion of the analogous muscle. Similarly, the term "tri-axle back plate" was historically used to describe the trapezius muscle.

The *superior* or *upper* (or descending) fibers of the trapezius originate from the spinous process of C7, the external occipital protuberance, the medial third of the superior nuchal line of the occipital bone (both in the back of the head), and the ligamentum nuchae. From this origin they proceed downward and laterally to be inserted into the posterior border of the lateral third of the clavicle.

The middle fibers, or transverse of the trapezius arise from the spinous process of the seventh cervical (both in the back of the neck), and the spinous processes of the first, second, and third thoracic vertebrae. They are inserted into the medial margin of the acromion, and into the superior lip of the posterior border of the spine of the scapula.

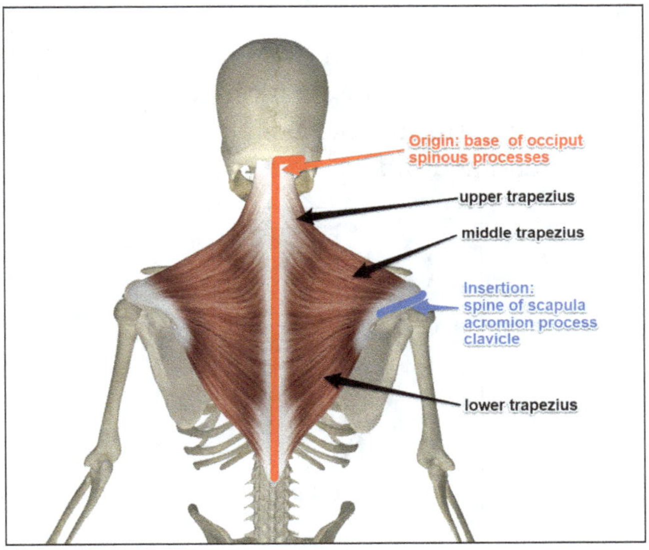

Trapezius muscle.

The *inferior* or *lower* (or ascending) fibers of the trapezius arise from the spinous processes of the remaining thoracic vertebrae (T4–T12). From this origin they proceed upward and laterally to converge near the scapula and end in an aponeurosis, which glides over the smooth triangular surface on the medial end of the spine, to be inserted into a tubercle at the apex of this smooth triangular surface.

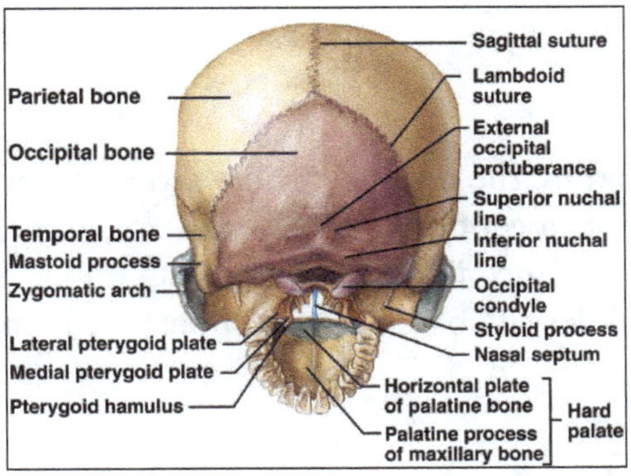

Occipital bone.

At its occipital origin, the trapezius is connected to the bone by a thin fibrous lamina, firmly adherent to the skin. The superficial and deep epimysia are continuous with an investing deep fascia that encircles the neck and also contains both sternocleidomastoid muscles.

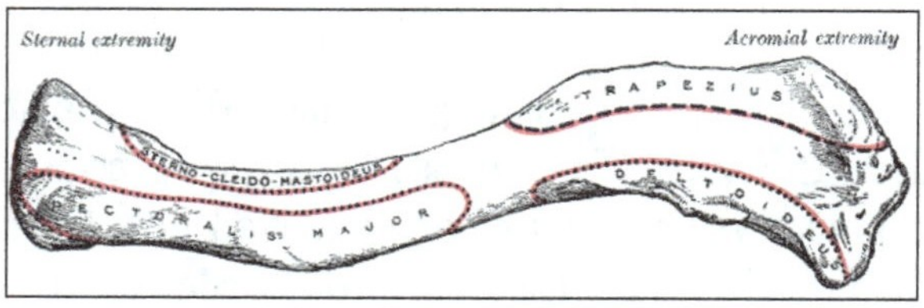

Left clavicle. Superior surface.

At the middle, the muscle is connected to the spinous processes by a broad semi-elliptical aponeurosis, which reaches from the sixth cervical to the third thoracic vertebræ and forms, with that of the opposite muscle, a tendinous ellipse. The rest of the muscle arises by numerous short tendinous fibers.

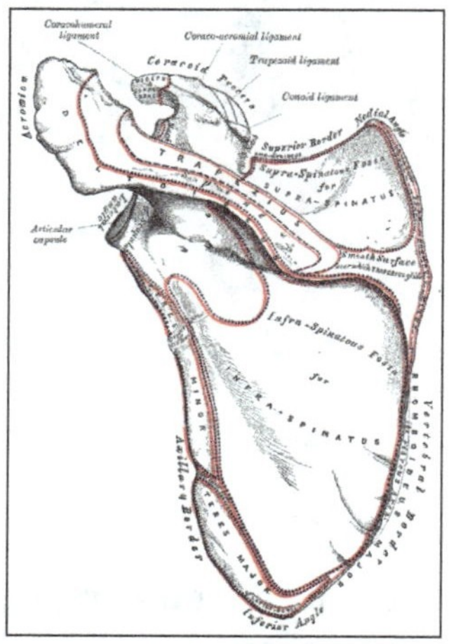

Left scapula. Posterior surface.

It is possible to feel the muscles of the superior trapezius become active by holding a weight in one hand in front of the body and, with the other hand, touching the area between the shoulder and the neck.

Nerve Supply

Motor function is supplied by the accessory nerve. Sensation, including pain and the sense of joint position (proprioception), travel via the ventral rami of the third (C3) and fourth (C4) cervical nerves. Since it is a muscle of the upper limb, the trapezius is not innervated by dorsal rami despite being placed superficially in the back.

Function

Contraction of the trapezius muscle can have two effects: movement of the scapulae when the spinal origins are stable, and movement of the spine when the scapulae are stable. Its main function is to stabilize and move the scapula.

Scapular Movements

The upper fibers elevate the scapulae, the middle fibers retract the scapulae, and the lower fibers depress the scapulae.

In addition to scapular translation, the trapezius induces scapular rotation. The upper and lower fibers tend to rotate the scapula around the Sternoclavicular articulation so that the acromion and inferior angles move up and the medial border moves down (upward rotation). The upper and lower fibers work in tandem with serratus anterior to upwardly rotate the scapulae, and work in opposition to the levator scapulae and the rhomboids, which effect downward rotation.

An example of trapezius function is an overhead press. When activating together, the upper and lower fibers also assist the middle fibers (along with other muscles such as the rhomboids) with scapular retraction/adduction.

The trapezius also assists in abduction of the shoulder above 90 degrees by rotating the glenoid upward. Injury to cranial nerve XI will cause weakness in abducting the shoulder above 90 degrees.

Spinal Movements

When the scapulae are stable a co-contraction of both sides can extend the neck.

Sternocleidomastoid Muscle

The sternocleidomastoid muscle is one of the largest and most superficial cervical muscles. The primary actions of the muscle are rotation of the head to the opposite side and flexion of the neck. The sternocleidomastoid is innervated by the accessory nerve.

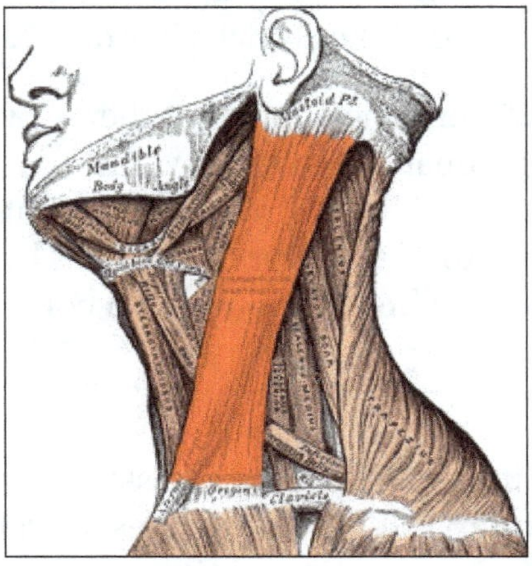

Neck muscles, with the sternocleidomastoid shaded.

It is given the name *sternocleidomastoid* because it originates at the manubrium of the sternum (*sterno-*) and the clavicle (*cleido-*), and has an insertion at the mastoid process of the temporal bone of the skull.

The sternocleidomastoid (right muscle shown) can be clearly observed when rotating the head.

Structure

The sternocleidomastoid muscle originates from two locations: the manubrium of the sternum and the clavicle. It travels obliquely across the side of the neck and inserts at the mastoid process of the temporal bone of the skull. The sternocleidomastoid is thick and narrow at its centre, and broader and thinner at either end.

The sternal head is a round fasciculus, tendinous in front, fleshy behind, arising from the upper part of the front of the manubrium sterni. It travels superiorly, laterally, and posteriorly.

The clavicular head is composed of fleshy and aponeurotic fibers, arises from the upper, frontal surface of the medial third of the clavicle; it is directed almost vertically upward.

The two heads are separated from one another at their origins by a triangular interval (supraclavicular fossa) but gradually blend, below the middle of the neck, into a thick, rounded muscle which is inserted, by a strong tendon, into the lateral surface of the mastoid process, from its apex to its superior border, and by a thin aponeurosis into the lateral half of the superior nuchal line of the occipital bone.

Nerve Supply

The sternocleidomastoid is innervated by the accessory nerve of the same side. It supplies only motor fibres. The cervical plexus supplies sensation, including proprioception, from the ventral primary rami of C2 and C3.

Variation

The clavicular origin of the sternocleidomastoid varies greatly: in some cases the clavicular head may be as narrow as the sternal; in others it may be as much as 7.5 millimetres (0.30 in) in breadth.

When the clavicular origin is broad, it is occasionally subdivided into several slips, separated by narrow intervals. More rarely, the adjoining margins of the sternocleidomastoid and trapezius are in contact. This would leave no posterior triangle.

The supraclavicularis muscle arises from the manubrium behind the sternocleidomastoid and passes behind the sternocleidomastoid to the upper surface of the clavicle.

Function

The function of this muscle is to rotate the head to the opposite side or obliquely rotate the head. It also flexes the neck. When both sides of the muscle act together, it flexes the neck and extends the head. When one side acts alone, it causes the head to rotate to the opposite side and flexes laterally to the same side (ipsilaterally).

It also acts as an accessory muscle of respiration, along with the scalene muscles of the neck.

Contraction

The signaling process to contract or relax the sternocleidomastoid begins in Cranial Nerve XI, the accessory nerve. The accessory nerve nucleus is in the anterior horn of the spinal cord around C1-C3, where lower motor neuron fibers mark its origin. The fibers from the accessory nerve nucleus travel upward to enter the cranium via the foramen magnum. The internal carotid artery to reach both the sternocleidomastoid muscles and the trapezius. After a signal reaches the accessory nerve nucleus in the anterior horn of the spinal cord, the signal is conveyed to motor endplates on the muscle fibers located at the clavicle. Acetylcholine (ACH) is released from vesicles and is sent over the synaptic cleft to receptors on the postsynaptic bulb. The ACH causes the resting potential to increase above -55mV, thus initiating an action potential which travels along the muscle fiber. Along the muscle fibers are t-tubule openings which facilitate the spread of the action potential into the muscle fibers. The t-tubule meets with the sarcoplasmic reticulum at locations throughout the muscle fiber, at these locations the sarcoplasmic reticulum releases calcium ions that results in the movement of troponin and tropomyosin on thin filaments. The movement of troponin and tropomyosin is key in facilitating the myosin head to move along the thin filament, resulting in a contraction of the sternocleidomastoid muscle.

Anatomical Landmark

The sternocleidomastoid is within the investing fascia of the neck, along with the trapezius muscle, with which it shares its nerve supply (the accessory nerve). It is thick and thus serves as a primary landmark of the neck, as it divides the neck into anterior and posterior cervical triangles (in front and behind the muscle, respectively) which helps define the location of structures, such as the lymph nodes for the head and neck.

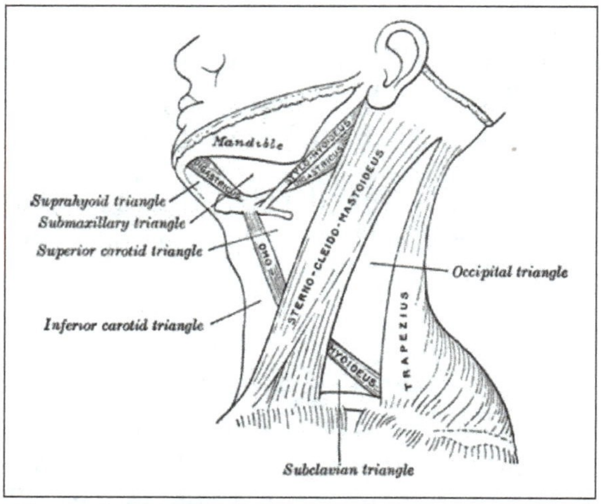

The sternocleidomastoid muscle with nearby structures labeled, such as the triangles of the neck.

Many important structures relate to the sternocleidomastoid, including the common carotid artery, accessory nerve, and brachial plexus.

Omohyoid Muscle

The omohyoid muscle is a muscle that depresses the hyoid. It is located in the front of the neck and consists of two bellies separated by an intermediate tendon. Its superior belly serves as the most lateral member of the infrahyoid muscles, located lateral to both the sternothyroid and thyrohyoid muscles. Its name derives from the Greek "omos" meaning shoulder, giving one of its attachments, and "hyoid", giving the other attachment – the hyoid bone.

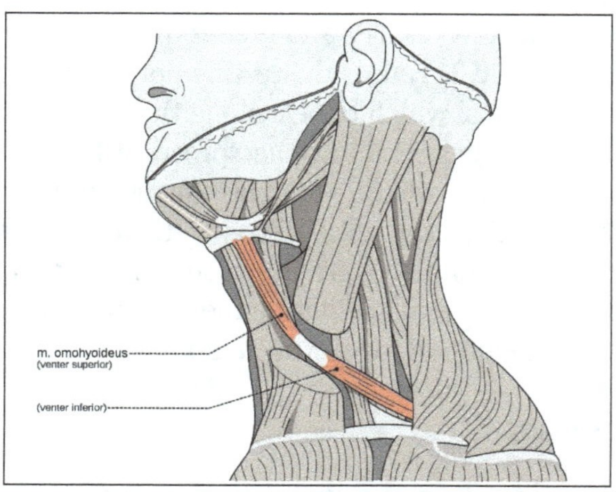

The omohyoid muscle, highlighted.

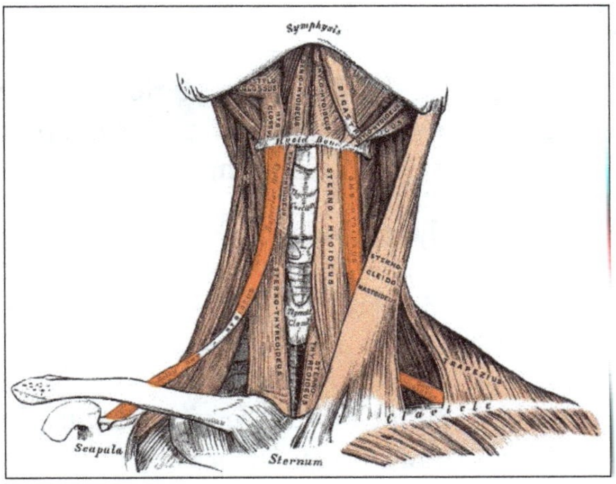

Muscles of the neck. Anterior view.

The omohyoid muscle is a muscle that depresses the hyoid. It is located in the front of the neck and consists of two bellies separated by an intermediate tendon. Its superior belly serves as the most lateral member of the infrahyoid muscles, located lateral to both the sternothyroid and thyrohyoid muscles.

Structure

It arises from the upper border of the scapula, and occasionally from the superior transverse scapular ligament which crosses the scapular notch, its extent of attachment to the scapula varying from a few millimetres to 2.5 cm.

This muscle has two separate bellies: superior and inferior.

From this origin, the *inferior belly* forms a flat, narrow fasciculus, which inclines forward and slightly upward across the lower part of the neck, being bound down to the clavicle by a fibrous expansion; it then passes behind the sternocleidomastoid, becomes tendinous and changes its direction, forming an obtuse angle.

It ends in the *superior belly*, which passes almost vertically upward, close to the lateral border of the sternohyoid, to be inserted into the lower border of the body of the hyoid bone, lateral to the insertion of the sternohyoid.

The central tendon of this muscle varies much in length and form, and is held in position by a process of the deep cervical fascia, which sheaths it, and is prolonged down to be attached to the clavicle and first rib; it is by this means that the angular form of the muscle is maintained. The tendon overlies the internal jugular vein, and can be used as a landmark for this vein during surgery.

Variation

Doubling; absence; origin from clavicle; absence or doubling of either belly.

Examination of the Neck

The inferior belly of the omohyoid divides the posterior triangle of the neck into an upper or occipital triangle and a lower or subclavian triangle. Its superior belly divides the anterior triangle into an upper or carotid triangle and a lower or muscular triangle.

The omohyoid muscle is proximally attached to the scapula and distally attached to the hyoid bone.

Innervation

The omohyoid is innervated by a branch of the cervical plexus, the ansa cervicalis, and mostly acts to stabilise the hyoid bone. The inferior belly of the omohyoid is innervated by the three cervical branches (C1-C3) that make up the ansa cervicalis, while the superior belly is innervated by the superior root of ansa cervicalis which contains only fibers from the first cervical spinal nerves (C1).

Rhomboid Major Muscle

The rhomboid major is a skeletal muscle on the back that connects the scapula with the vertebrae of the spinal column. In human anatomy, it acts together with the rhomboid minor to keep the scapula pressed against thoracic wall and to retract the scapula toward the vertebral column.

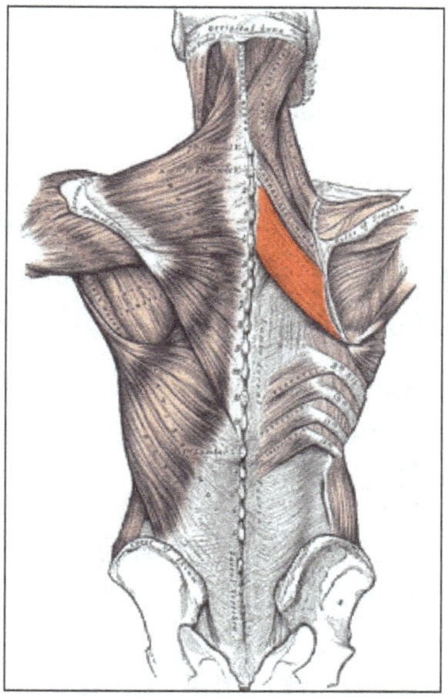

Muscles connecting the upper extremity to the vertebral column. Rhomboid major indicated in red.

Structure

The rhomboid major arises from the spinous processes of the thoracic vertebrae T2 to

T5 as well as the supraspinous ligament. It inserts on the medial border of the scapula, from about the level of the scapular spine to the scapula's inferior angle.

The rhomboid major is considered a superficial back muscle. It is deep to the trapezius, and is located directly inferior to the rhomboid minor. As the word *rhomboid* suggests, the rhomboid major is diamond-shaped. The *major* in its name indicates that it is the larger of the two rhomboids.

Variation

The two rhomboids are sometimes fused into a single muscle.

Nerve Supply

The rhomboid major, like the rhomboid minor, is innervated by the ventral primary ramus via the dorsal scapular nerve (C5).

Blood Supply

Both rhomboid muscles also derive their arterial blood supply from the dorsal scapular artery.

Function

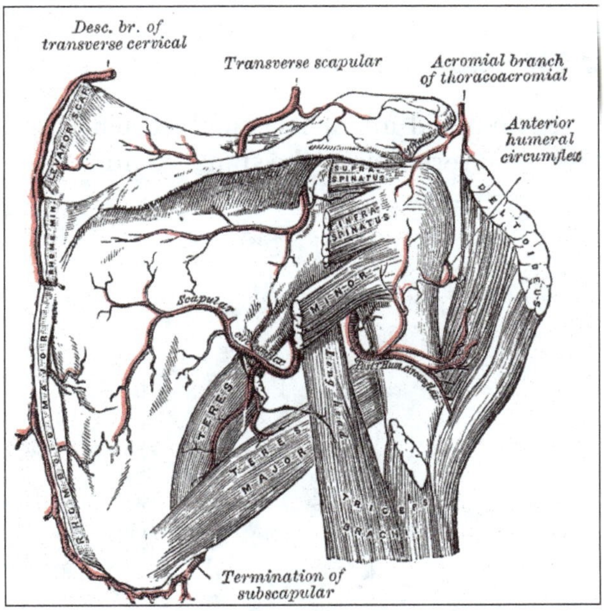

The scapular and circumflex arteries. Rhomboid major labelled at lower left.

The rhomboid major helps to hold the scapula (and thus the upper limb) onto the rib-cage. Other muscles that perform this function include the serratus anterior and pectoralis minor.

Both rhomboids (major and minor) also act to retract the scapula, pulling it towards the vertebral column.

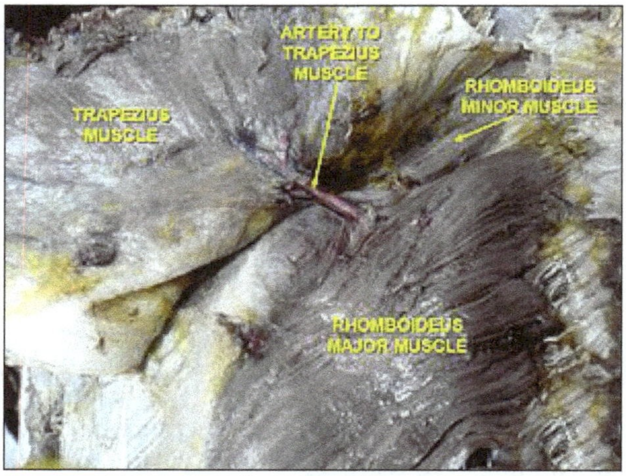

Rhomboideus major muscle.

The rhomboids work collectively with the levator scapulae muscles to elevate the medial border of the scapula, downwardly rotating the scapula with respect to the glenohumeral joint. Antagonists to this function (upward rotators of the scapulae) are the serratus anterior and lower fibers of the trapezius. If the lower fibers are inactive, the serratus anterior and upper trapezius work in tandem with rhomboids and levators to elevate the entire scapula.

Rhomboid Minor Muscle

In human anatomy, the rhomboid minor is a small skeletal muscle on the back that connects the scapula with the vertebrae of the spinal column.

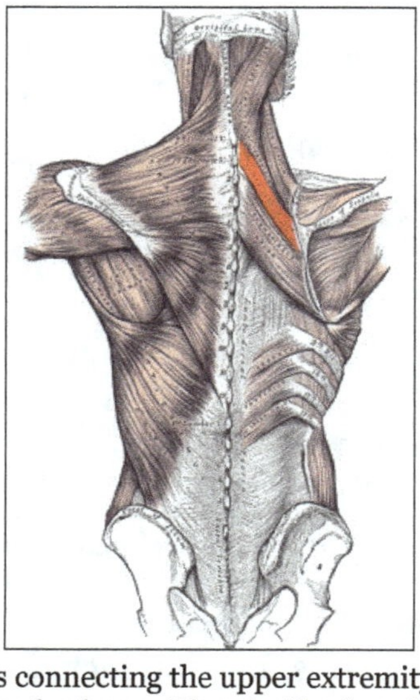

Muscles connecting the upper extremity to the vertebral column. Rhomboid minor in red.

Located inferior to levator scapulae and superior to rhomboid major, it acts together with the latter to keep the scapula pressed against the thoracic wall. It lies deep to trapezius but superficial to the long spinal muscles.

The rhomboid minor arises from the inferior border of the nuchal ligament, from the spinous processes of the seventh cervical and first thoracic vertebrae, and from the intervening supraspinous ligaments.

It is inserted into a small area of the medial border of the scapula at the level of the scapular spine.

Action

Together with the rhomboid major, the rhomboid minor retracts the scapula when trapezius is contracted. Acting as a synergist to the trapezius, the rhomboid major and minor elevate the medial border of the scapula medially and upward, working in tandem with the levator scapulae muscle to rotate the scapulae downward. While other shoulder muscles are active, the rhomboid major and minor stabilize the scapula.

Innervation and Blood Supply

The nerve supply comes from the dorsal scapular nerve, with most of its fibers derived from the C5 nerve root and only minor contribution from C4 or C6. The rhomboid minor gets its arterial blood supply from the dorsal scapular artery.

Variation

It is usually separated from the rhomboid major by a slight interval, but the adjacent margins of the two muscles are occasionally united.

Levator Scapulae Muscle

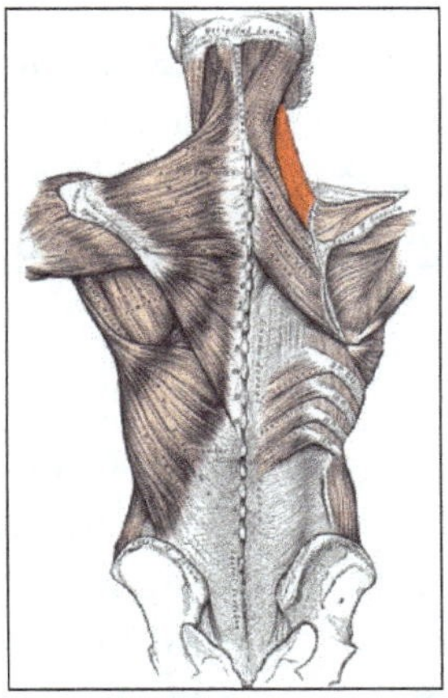

Muscles connecting the upper extremity to the vertebral column.
(Levator scapulae visible at upper right, at the neck).

The levator scapulae is a skeletal muscle situated at the back and side of the neck.

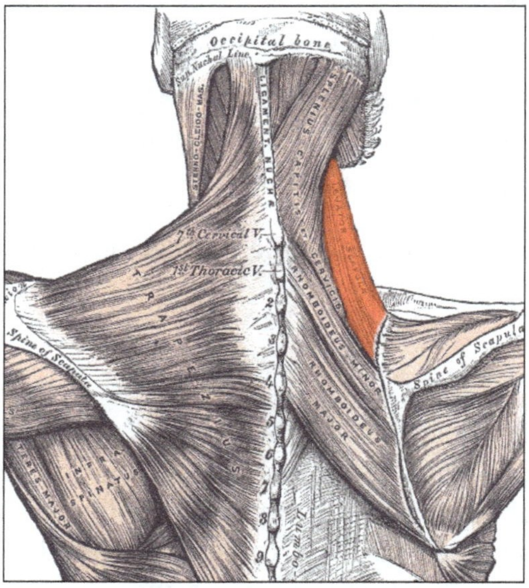

Muscles of neck. Levator scapulae visible at center left.

Structure

The levator scapulae originates from the posterior tubercle of the transverse process of cervical vertebrae one to four. The muscle is inserted into medial border of the scapula extending from superior angle to junction of spine and medial border of scapula.

The levator scapulae may lie deep to the Sternocleidomastoid at its origin, deep or adjacent to the splenius capitis at its origin and mid-portion, and deep to the trapezius in its lower portion.

Relations

One of the muscles within the floor of the posterior triangle of the neck, the superior part of levator scapulae is covered by sternocleidomastoid and its inferior part by the trapezius. It is bounded in front by the scalenus medius and behind by splenius cervicis. The spinal accessory nerve crosses laterally in the middle part of the muscle and the dorsal scapular nerve may lie deep to or pass through it.

Variation

The number of attachments varies; a slip may extend to the occipital or mastoid, to the trapezius, scalene or serratus anterior, or to the first or second rib. The muscle may be subdivided into several distinct parts from origin to insertion. Levator claviculæ from the transverse processes of one or two upper cervical vertebræ to the outer end of the clavicle corresponds to a muscle of lower animals. More or less union with the serratus anterior muscle.

Nerve Supply

The levator scapulae is supplied by two or three branches of the fourth and fifth cervical nerves, and frequently by a branch from the dorsal scapular nerve.

Blood Supply

The levator scapulae is supplied by the dorsal scapular artery. Normally, this artery has a small branch which passes laterally to the supraspinatus fossa of the scapula, and in a third of cases, this branch supplies the muscle. If the dorsal scapular artery comes off the transverse cervical artery, the parent transverse cervical artery splits, the dorsal scapular artery passes medially, while the transverse cervical artery passes laterally.

Function

When the spine is fixed, levator scapulae elevates the scapula and rotates its inferior angle medially. It often works in combination with other muscles like the rhomboids and pectoralis minor to rotate down.

Elevating or rotating one shoulder at a time would require muscles to stabilize the cervical spine and keep it immobile so it does not flex or rotate. Elevating both at once with equal amounts of pull on both side of cervical spinal origins would counteract these forces. Downward rotation would be prevented by co-contraction of other muscles that elevate the spine, the upper fibers of the trapezius, which is an upward rotator.

When the shoulder is fixed, levator scapulae rotates to the same side and flexes the cervical spine laterally. When both shoulders are fixed, a simultaneous co-contraction of both levator scapulae muscles in equal amounts would not produce lateral flexion or rotation, and may produce straight flexion or extension of the cervical spine.

Subclavius Muscle

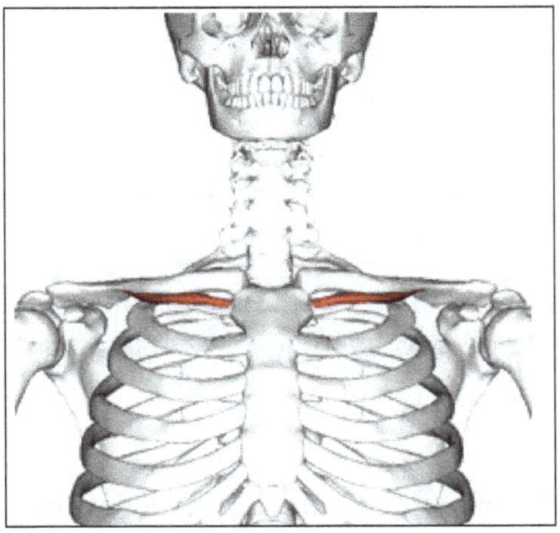

Subclavius muscle (shown in red).

The subclavius is a small triangular muscle, placed between the clavicle and the first rib. Along with the pectoralis major and pectoralis minor muscles, the subclavius muscle makes up the *Anterior Axioappendicular Muscles* also known as anterior wall of the axilla.

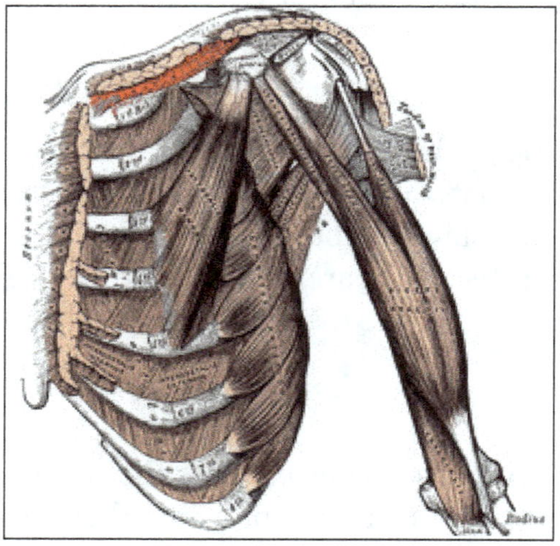

Deep muscles of the chest and front of the arm, with the boundaries of the axilla. (Subclavius visible at upper left, above first rib.)

Structure

It arises by a short, thick tendon from the first rib and its cartilage at their junction, in front of the costoclavicular ligament.

The fleshy fibers proceed obliquely superolaterally, to be inserted into the groove on the under surface of the clavicle.

Innervation

The nerve to subclavius (or subclavian nerve) innervates the muscle. This arises from the junction of the fifth and sixth cervical nerves, from the Superior/upper trunk of the brachial plexus.

Variation

Insertion into coracoid process instead of clavicle or into both clavicle and coracoid process.

Sternoscapular fasciculus to the upper border of scapula. Sternoclavicularis from manubrium to clavicle between pectoralis major and coracoclavicular fascia.

Function

The subclavius depresses the shoulder, carrying it downward and forward. It draws the clavicle inferiorly as well as anteriorly.

The subclavius protects the underlying brachial plexus and subclavian vessels from a broken clavicle - the most frequently broken long bone.

Pectoralis Minor

The pectoralis minor is a thin, triangular muscle, situated at the upper part of the chest, beneath the pectoralis major in the human body.

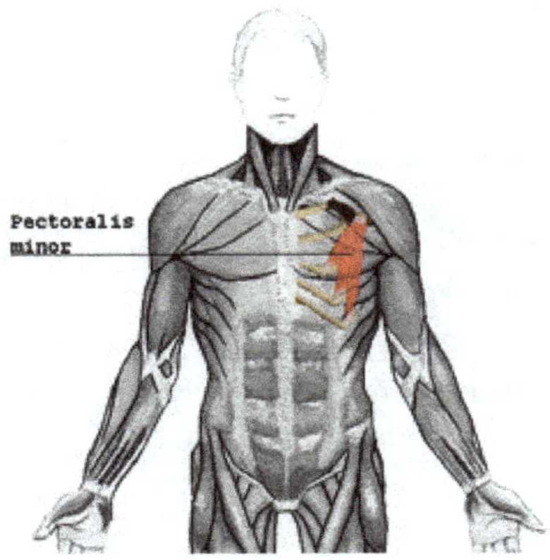	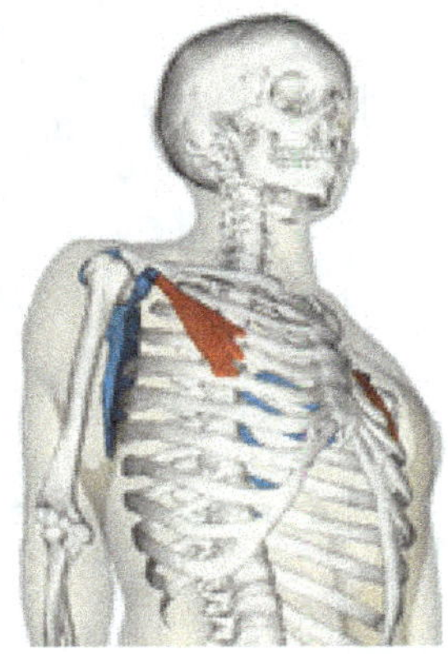
Pectoralis minor (shown in red).	Pectoralis minor muscle (shown in red). The bone shown in blue is the shoulder blade.

Structure

It arises from the upper margins and outer surfaces of the third, fourth, and fifth ribs, near their cartilages and from the aponeuroses covering the intercostalis.

The fibers pass superior and lateral and converge to form a flat tendon, which is inserted into the medial border and upper surface of the coracoid process of the scapula.

Relations

The pectoralis minor muscle is covered anteriorly (superficially) by the clavipectoral fascia. The Medial pectoral nerve pierces the pectoralis minor and the clavipectoral fascia. In attaching to the coracoid process, the pectoralis minor forms a 'bridge' - structures passing into the upper limb from the thorax will pass directly underneath.

Axillary nodes are classified according to their positions relative to the pectoralis minor muscle. Level 1 are lateral, Level 2 are deep, Level 3 are medial. The pectoralis minor divides the axillary artery into three parts (in contrary sequence compared to the nodes) - first part medial, second part deep/posterior, third part lateral in relation to the pectoralis minor.

Variations

The origin is from the second, third and fourth or fifth ribs. The tendon of insertion may extend over the coracoid process to the greater tubercle. It may be split into several parts. Absence of this muscle is rare but happens with certain uncommon diseases, such as the Poland syndrome.

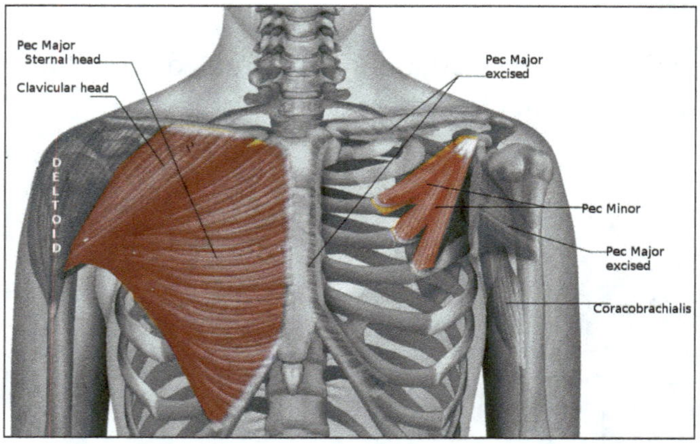

An individual with an axillary arch as well as sternalis muscle and an accessory abdominal portion of the pectoralis major.

Function

The pectoralis minor depresses the point of the shoulder, drawing the scapula superior, towards the thorax, and throwing its inferior angle posteriorly.

Serratus Anterior Muscle

The serratus anterior is a muscle that originates on the surface of the 1st to 8th ribs at the side of the chest and inserts along the entire anterior length of the medial border of the scapula. The serratus anterior acts to pull the scapula forward around the thorax.

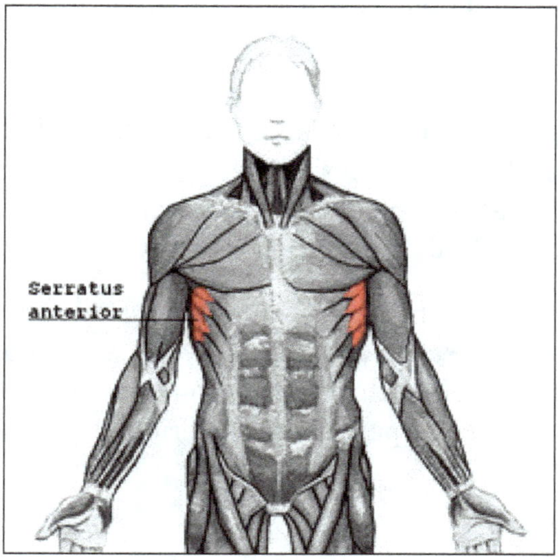

Serratus anterior, showing origin from lower ribs (origin from upper ribs obscured by pectoralis major and other superficial muscles).

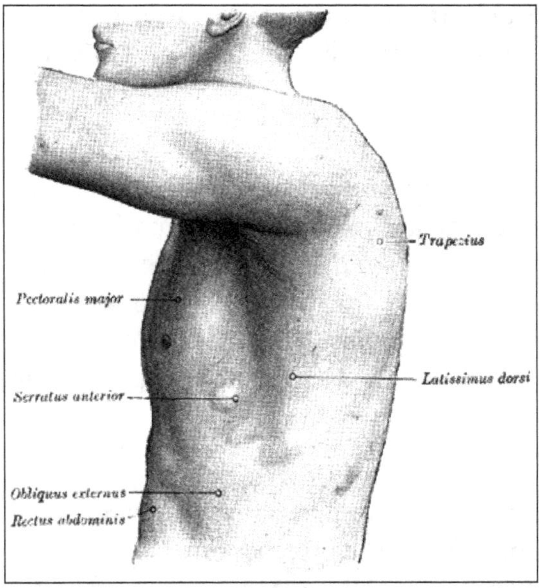

The left side of the thorax.

Structure

Serratus anterior normally originates by nine or ten slips (muscle branches) from either the first to ninth ribs or the first to eighth ribs. Because two slips usually arise from the second rib, the number of slips is greater than the number of ribs from which they originate.

The muscle is inserted along the medial border of the scapula between the superior and inferior angles along with being inserted along the thoracic vertebrae. The muscle is divided into three named parts depending on their points of insertions:

- The serratus anterior superior is inserted near the superior angle.
- The serratus anterior intermediate is inserted along the medial border.
- The serratus anterior inferior is inserted near the inferior angle.

Relations

The serratus anterior lies deep to the subscapularis, from which it is separated by the subscapularis (supraserratus) bursa. It is separated from the rib by the scapulothoracic (infraserratus) bursa.

Nerve Supply

The serratus anterior is innervated by the long thoracic nerve (Nerve of Bell), a branch of the brachial plexus. The long thoracic nerve travels inferiorly on the surface of the serratus. The nerve is especially vulnerable during certain types of surgery (for example, during lymph node clearance from the axilla (e.g., in case of axillary dissection in a surgery for breast cancer)). Damage to this nerve can lead to a winged scapula.

Function

All three parts described above pull the scapula forward around the thorax, which is essential for anteversion of the arm. As such, the muscle is an antagonist to the rhomboids. However, when the inferior and superior parts act together, they keep the scapula pressed against the thorax *together* with the rhomboids and therefore these parts also act as synergists to the rhomboids. The inferior part can pull the lower end of the scapula laterally and forward and thus rotates the scapula to make elevation of the arm possible. Additionally, all three parts can lift the ribs when the shoulder girdle is fixed, and thus assist in respiration.

The serratus anterior is occasionally called the "big swing muscle" or "boxer's muscle" because it is largely responsible for the protraction of the scapula — that is, the pulling of the scapula forward and around the rib cage that occurs when someone throws a punch.

The serratus anterior also plays an important role in the upward rotation of the scapula, such as when lifting a weight overhead. It performs this in sync with the upper and lower fibers of the trapezius.

Splenius Capitis Muscle

The splenius capitis is a broad, straplike muscle in the back of the neck. It pulls on the base of the skull from the vertebrae in the neck and upper thorax. It is involved in movements such as shaking the head.

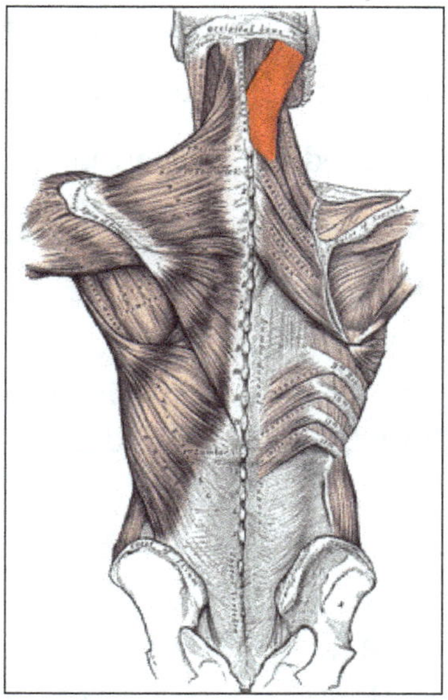

Muscles connecting the upper extremity to the vertebral column (splenius capitis et cervicis labeled at upper right).

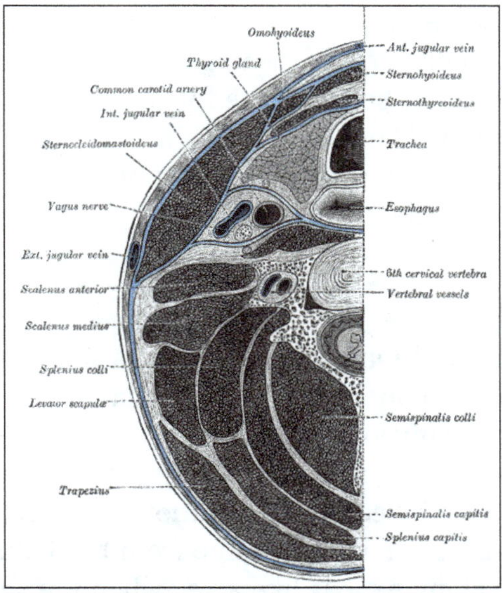

Section of the neck at about the level of the sixth cervical vertebra. Showing the arrangement of the deep cervical fascia (splenius capitis labeled at bottom right).

Structure

It arises from the lower half of the nuchal ligament, from the spinous process of the seventh cervical vertebra, and from the spinous processes of the upper three or four thoracic vertebrae.

The fibers of the muscle are directed upward and laterally and are inserted, under cover of the sternocleidomastoideus, into the mastoid process of the temporal bone, and into the rough surface on the occipital bone just below the lateral third of the superior nuchal line. The splenius capitis is deep to sternocleidomastoideus at the mastoid process, and to the trapezius for its lower portion. It is one of the muscles that forms the floor of the posterior triangle of the neck.

The splenius capitis muscle is innervated by the posterior ramus of spinal nerves C3 and C4.

Function

The splenius capitis muscle is a prime mover for head extension. The splenius capitis can also allow lateral flexion and rotation of the cervical spine.

Bones of the Upper Limb

The upper limb is divided into three regions. These consist of the arm, located between the shoulder and elbow joints; the forearm, which is between the elbow and wrist joints;

and the hand, which is located distal to the wrist. There are 30 bones in each upper limb. The humerus is the single bone of the upper arm, and the ulna (medially) and the radius (laterally) are the paired bones of the forearm. The base of the hand contains eight bones, each called a carpal bone, and the palm of the hand is formed by five bones, each called a metacarpal bone. The fingers and thumb contain a total of 14 bones, each of which is a phalanx bone of the hand.

Clavicle

The clavicle, or collarbone, is a long bone that serves as a strut between the shoulder blade and the sternum (breastbone). There are two clavicles, one on the left and one on the right. The clavicle is the only long bone in the body that lies horizontally. Together with the shoulder blade, it makes up the shoulder girdle. It is a touchable bone, and in people who have less fat in this region, the location of the bone is clearly visible, as it creates a bulge in the skin. It receives its name from the Latin clavicula ("little key"), because the bone rotates along its axis like a key when the shoulder is abducted. The clavicle is the most commonly fractured bone. It can easily be fractured due to impacts to the shoulder from the force of falling on outstretched arms or by a direct hit.

Structure

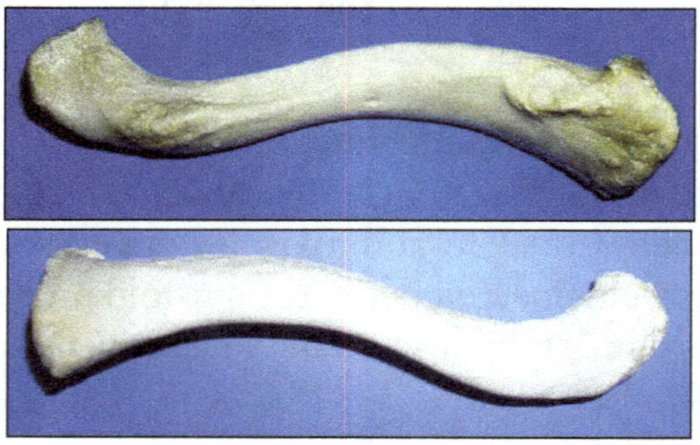

Right clavicle—from below, and from above.

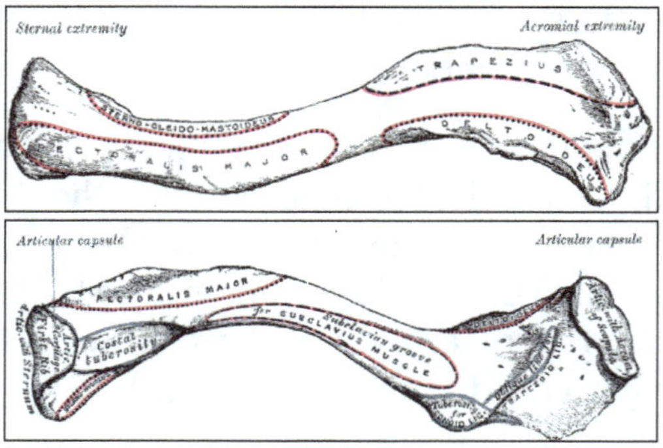

Left clavicle—from above, and from below.

The collarbone is a large doubly curved long bone that connects the arm to the trunk of the body. Located directly above the first rib, it acts as a strut to keep the scapula in place so that the arm can hang freely. Medially, it articulates with the manubrium of the sternum (breastbone) at the sternoclavicular joint. At its lateral end it articulates with the acromion, a process of the scapula (shoulder blade), at the acromioclavicular joint. It has a rounded medial end and a flattened lateral end.

From the roughly pyramidal sternal end, each collarbone curves laterally and anteriorly for roughly half its length. It then forms an even larger posterior curve to articulate with the acromion of the scapula. The flat acromial end of the collarbone is broader than the sternal end. The acromial end has a rough inferior surface that bears a ridge, the trapezoid line, and a slight rounded projection, the conoid tubercle (above the coracoid process). These surface features are attachment sites for muscles and ligaments of the shoulder.

It can be divided into three parts: medial end, lateral end and shaft.

Medial End

It is also known as the sternal end. The medial end is quadrangular and articulates with the clavicular notch of the manubrium of the sternum to form the sternoclavicular joint. The articular surface extends to the inferior aspect for attachment with the first costal cartilage.

It gives attachments to:

- Fibrous capsule joint, all around.
- Articular disc, superoposteriorly.
- Interclavicular ligament, superiorly.

Lateral End

The lateral end is also known as the acromial end. It is flat from above downward. It bears a facet that articulates with the shoulder to form the acromioclavicular joint. The area surrounding the joint gives an attachment to the joint capsule. The anterior border is concave forward and posterior border is convex backward.

Shaft

The shaft is divided into the medial two-thirds and the lateral one third. The medial part is thicker than the lateral.

Lateral One-third of the Shaft

The lateral third of the shaft has two borders and two surfaces.

- The anterior border is concave forward and gives origin to the deltoid muscle.
- The posterior border is convex and gives attachment to the trapezius muscle.
- The inferior surface has a ridge called the trapezoid line and a tubercle; the conoid tubercle for attachment with the trapezoid and the conoid ligament, part of the coracoclavicular ligament that serves to connect the collarbone with the coracoid process of the scapula.

Development

The collarbone is the first bone to begin the process of ossification (laying down of minerals onto a preformed matrix) during development of the embryo, during the fifth and sixth weeks of gestation. However, it is one of the last bones to finish ossification at about 21–25 years of age. Its lateral end is formed by intramembranous ossification while medially it is formed by endochondral ossification. It consists of a mass of cancellous bone surrounded by a compact bone shell. The cancellous bone forms via two ossification centres, one medial and one lateral, which fuse later on. The compact forms as the layer of fascia covering the bone stimulates the ossification of adjacent tissue. The resulting compact bone is known as a periosteal collar.

Even though it is classified as a long bone, the collarbone has no medullary (bone marrow) cavity like other long bones, though this is not always true. It is made up of spongy cancellous bone with a shell of compact bone. It is a dermal bone derived from elements originally attached to the skull.

Variation

The shape of the clavicle varies more than most other long bones. It is occasionally pierced by a branch of the supraclavicular nerve. In males the clavicle is generally longer and more massive than in females. A study measuring 748 males and 252 females saw a difference in collarbone length between age groups 18–20 and 21–25 of about 6 and 5 mm (0.24 and 0.20 in) for males and females respectively.

The left clavicle is usually longer and not as strong as the right clavicle. The collarbones are sometimes partly or completely absent in cleidocranial dysostosis. The levator claviculae muscle, present in 2–3% of people, originates on the transverse processes of the upper cervical vertebrae and is inserted in the lateral half of the clavicle.

Functions

The collarbone serves several functions:

- It serves as a rigid support from which the scapula and free limb suspended; an arrangement that keeps the upper limb away from the thorax so that the arm has maximum range of movement. Acting as a flexible, crane-like strut, it allows the scapula to move freely on the thoracic wall.
- Covering the cervicoaxillary canal, it protects the neurovascular bundle that supplies the upper limb.
- Transmits physical impacts from the upper limb to the axial skeleton.

Muscle

Muscles and ligaments that attach to the collarbone include:

Attachment on collarbone	Muscle/Ligament	Other attachment
Superior surface and anterior border	Deltoid muscle	Deltoid tubercle, anteriorly on the lateral third
Superior surface	Trapezius muscle	Posteriorly on the lateral third
Inferior surface	Subclavius muscle	Subclavian groove
Inferior surface	Conoid ligament (the medial part of the coracoclavicular ligament)	Conoid tubercle
Inferior surface	Trapezoid ligament (the lateral part of the coracoclavicular ligament)	Trapezoid line
Anterior border	Pectoralis major muscle	Medial third (rounded border)
Posterior border	Sternocleidomastoid muscle (clavicular head)	Superiorly, on the medial third
Posterior border	Sternohyoid muscle	Inferiorly, on the medial third
Posterior border	Trapezius muscle	Lateral third

Clinical Significance

- Acromioclavicular dislocation ("AC Separation").
- Degeneration of the clavicle.
- Osteolysis.
- Sternoclavicular dislocations.

A vertical line drawn from the mid-clavicle called the mid-clavicular line is used as a reference in describing cardiac apex beat during medical examination. It is also useful for evaluating an enlarged liver, and for locating the gallbladder which is between the mid-clavicular line and the transpyloric plane.

Collarbone Fracture

Clavicle fractures (colloquially, a broken collarbone) occur as a result of injury or trauma. The most common type of fractures occur when a person falls horizontally on the shoulder or with an outstretched hand. A direct hit to the collarbone will also cause a break. In most cases, the direct hit occurs from the lateral side towards the medial side of the bone. The most common site of fracture is the junction between the two curvatures of the bone, which is the weakest point. This results in the sternocleidomastoid muscle lifting the medial aspect superiorly, which can result in perforation of the overlying skin.

Scapula

In anatomy, the scapula (plural scapulae or scapulas), also known as the shoulder bone, shoulder blade, wing bone or blade bone, is the bone that connects the humerus (upper arm bone) with the clavicle (collar bone). Like their connected bones, the scapulae are paired, with each scapula on either side of the body being roughly a mirror image of the other.

The scapula forms the back of the shoulder girdle. In humans, it is a flat bone, roughly triangular in shape, placed on a posterolateral aspect of the thoracic cage.

Structure

The scapula is a wide, flat bone lying on the thoracic wall that provides an attachment for three groups of muscles: intrinsic, extrinsic, and stabilising and rotating muscles. The intrinsic muscles of the scapula include the muscles of the rotator cuff—the subscapularis, teres minor, supraspinatus, and infraspinatus. These muscles attach to the surface of the scapula and are responsible for the internal and external rotation of the shoulder joint, along with humeral abduction.

The extrinsic muscles include the biceps, triceps, and deltoid muscles and attach to the coracoid process and supraglenoid tubercle of the scapula, infraglenoid tubercle of the scapula, and spine of the scapula. These muscles are responsible for several actions of the glenohumeral joint.

The third group, which is mainly responsible for stabilization and rotation of the scapula, consists of the trapezius, serratus anterior, levator scapulae, and rhomboid muscles. These attach to the medial, superior, and inferior borders of the scapula.

The head, processes, and the thickened parts of the bone contain cancellous tissue; the rest consists of a thin layer of compact tissue. The central part of the supraspinatus fossa and the upper part of the infraspinatous fossa, but especially the former, are usually so thin in humans as to be semitransparent; occasionally the bone is found wanting in this situation, and the adjacent muscles are separated only by fibrous tissue. The scapula has two surfaces, three borders, three angles, and three processes.

Surfaces

The front of the scapula (also known as the costal or ventral surface) has a broad concavity called the subscapular fossa, to which the subscapularis muscle attaches. The medial two-thirds of the fossa have 3 longitudinal oblique ridges, and another thick ridge adjoins the lateral border; they run outward and upward. The ridges give attachment to the tendinous insertions, and the surfaces between them to the fleshy fibers, of the subscapularis muscle. The lateral third of the fossa is smooth and covered by the fibers of this muscle.

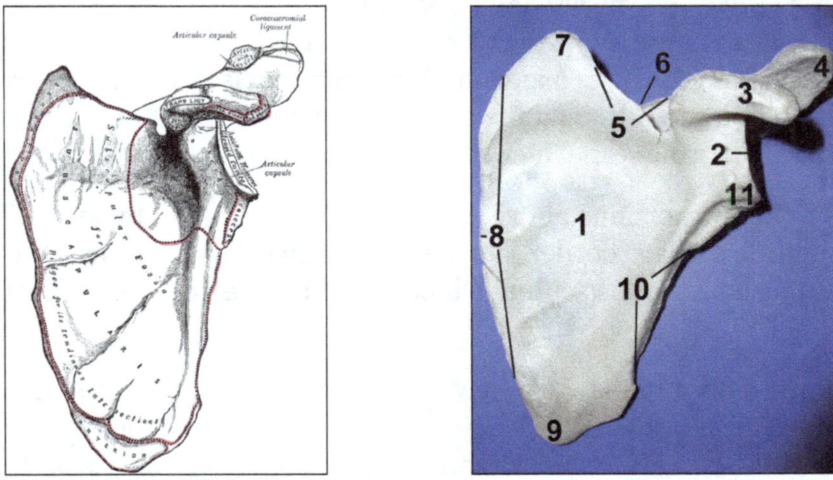

Left scapula: Costal surface: (1) Subscapular fossa (2) Glenoid cavity (3) Coracoid process (4) Acromion (5) Superior border (6) Scapular notch (7) Superior angle (8) Medial border (9) Inferior angle (10) Lateral border+ (11) Infraglenoid tubercle.

At the upper part of the fossa is a transverse depression, where the bone appears to be bent on itself along a line at right angles to and passing through the center of the glenoid cavity, forming a considerable angle, called the subscapular angle; this gives greater strength to the body of the bone by its arched form, while the summit of the arch serves to support the spine and acromion.

The costal surface superior of the scapula is the origin of 1st digitation for the serratus anterior origin.

Back

The back of the scapula (also called the dorsal or posterior surface) is arched from above downward, and is subdivided into two unequal parts by the spine of the scapula. The portion above the spine is called the supraspinous fossa, and that below it the infraspinous fossa. The two fossae are connected by the spinoglenoid notch, situated lateral to the root of the spine.

- The supraspinous fossa, the smaller of the two, is concave, smooth, and broader at its vertebral than at its humeral end; its medial two-thirds give origin to the Supraspinatus.
- The infraspinous fossa is much larger than the preceding; toward its vertebral margin a shallow concavity is seen at its upper part; its center presents a prominent convexity, while near the axillary border is a deep groove which runs from the upper toward the lower part. The medial two-thirds of the fossa give origin to the Infraspinatus; the lateral third is covered by this muscle.

There is a ridge on the outer part of the back of the scapula. This runs from the lower part of the glenoid cavity, downward and backward to the vertebral border, about 2.5 cm above the inferior angle. Attached to the ridge is a fibrous septum, which separates the infraspinatus muscle from the Teres major and Teres minor muscles. The upper two-thirds of the surface between the ridge and the axillary border is narrow, and is crossed near its center by a groove for the scapular circumflex vessels; the Teres minor attaches here.

The broad and narrow portions above alluded to are separated by an oblique line, which runs from the axillary border, downward and backward, to meet the elevated ridge: to it is attached a fibrous septum which separates the Teres muscles from each other.

Its lower third presents a broader, somewhat triangular surface, the inferior angle of the scapula, which gives origin to the Teres major, and over which the Latissimus dorsi glides; frequently the latter muscle takes origin by a few fibers from this part.

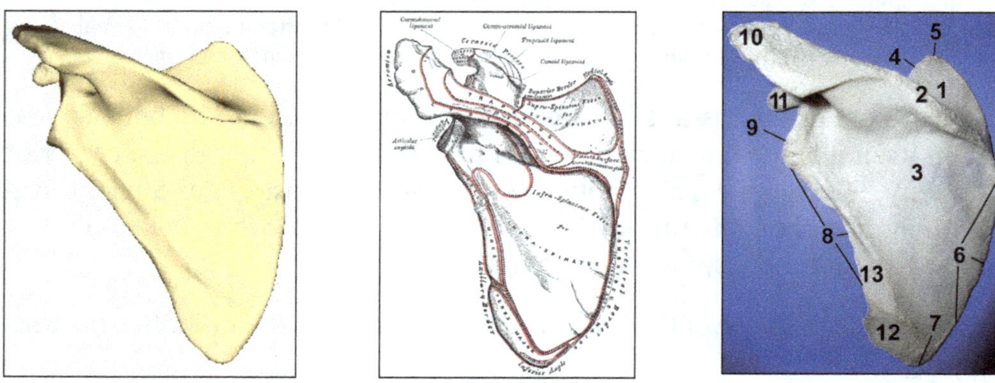

Left scapula: Dorsal surface: (1) Supraspinous fossa (2) Spine (3) Infraspinous fossa (4) Superior border (5) Superior angle (6) Medial border (7) Inferior angle (8) Lateral border (9) Lateral angle (10) Acromion (11) Coracoid process (12) Origin of teres major muscle (13) Origin of teres minor muscle.

Side

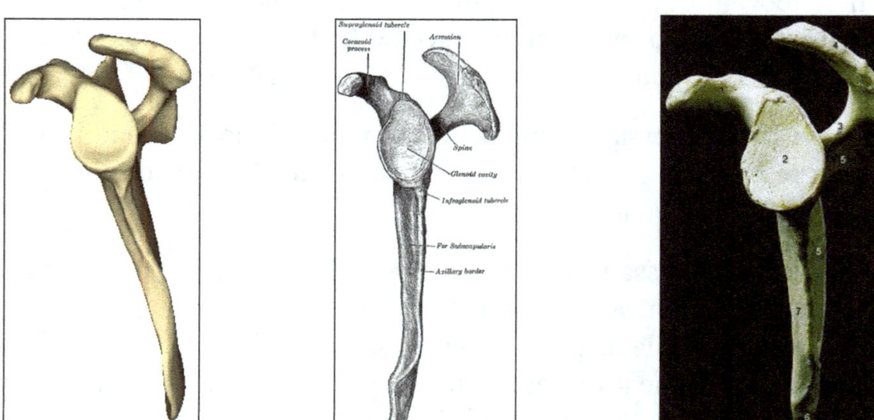

Left scapula: Lateral surface: (1) Coracoid process (2) Glenoid cavity (3) Supraspinous fossa (4) Acromion (5) Infraspinous fossa (6) Inferior angle (7) Lateral border.

The acromion forms the summit of the shoulder, and is a large, somewhat triangular or oblong process, flattened from behind forward, projecting at first laterally, and then curving forward and upward, so as to overhang the glenoid cavity.

Angles

There are 3 angles:

- The superior angle of the scapula or medial angle, is covered by the trapezius muscle. This angle is formed by the junction of the superior and medial borders of the scapula. The superior angle is located at the approximate level of the second thoracic vertebra. The superior angle of the scapula is thin, smooth, rounded, and inclined somewhat lateralward, and gives attachment to a few fibers of the levator scapulae muscle.
- The inferior angle of the scapula is the lowest part of the scapula and is covered by the latissimus dorsi muscle. It moves forwards round the chest when the arm is abducted. The inferior angle is formed by the union of the medial and lateral borders of the scapula. It is thick and rough and its posterior or back surface affords attachment to the teres major and often to a few fibers of the latissimus dorsi. The anatomical plane that passes vertically through the inferior angle is named the scapular line.
- The lateral angle of the scapula or glenoid angle also known as the head of the scapula is the thickest part of the scapula. It is broad and bears the glenoid cavity on its articular surface which is directed forward, laterally and slightly upwards, and articulates with the head of the humerus. The inferior angle is broader below than above and its vertical diameter is the longest. The surface is covered with cartilage in the fresh state; and its margins, slightly raised, give attachment to a fibrocartilaginous structure, the glenoidal labrum, which deepens the cavity. At its apex is a slight elevation, the supraglenoid tuberosity, to which the long head of the biceps brachii is attached.

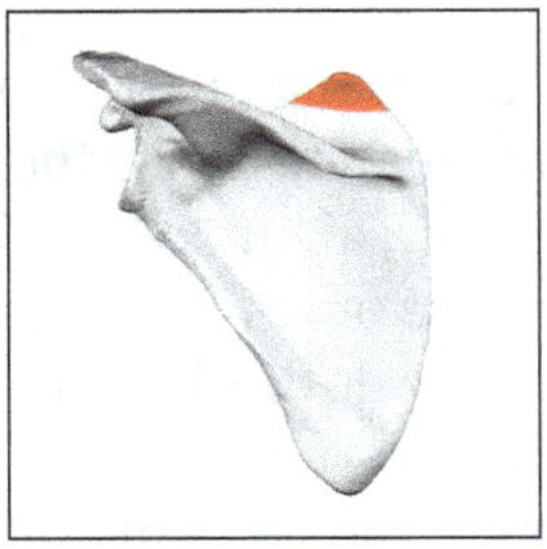

Superior angle shown in red.

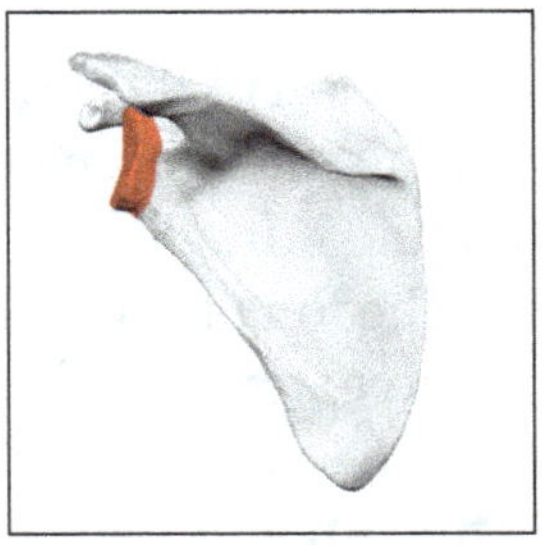

Lateral angle shown in red.

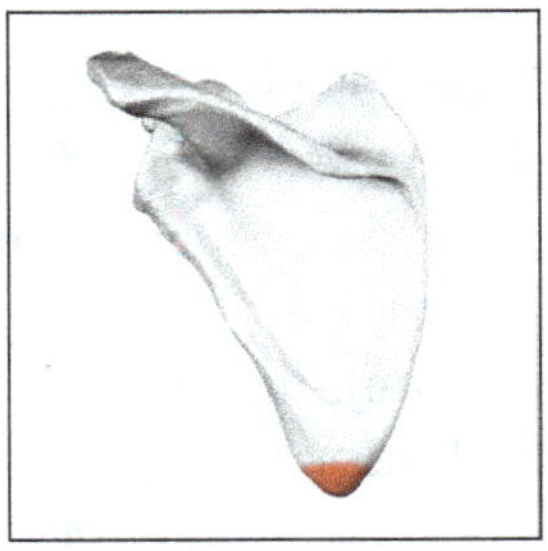

Inferior angle shown in red.

The anatomic neck of the scapula is the slightly constricted portion which surrounds the head and is more distinct below and behind than above and in front. The surgical neck of the scapula passes directly medial to the base of the coracoid process.

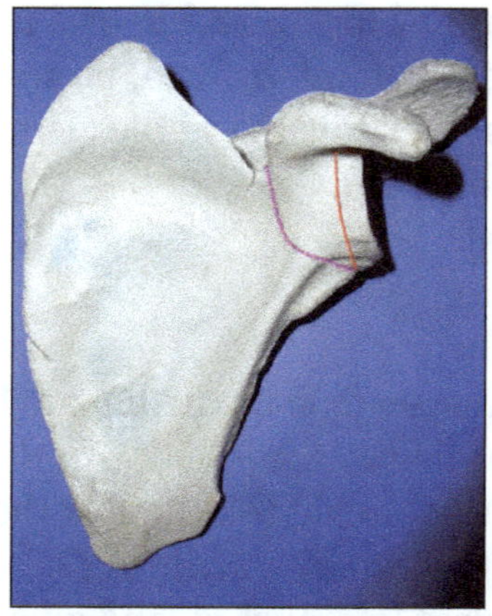

Anatomic neck: red, Surgical neck: purple.

Borders

There are three borders of the scapula:

- The superior border is the shortest and thinnest; it is concave, and extends from the superior angle to the base of the coracoid process. It is referred to as the cranial border in animals.
    - At its lateral part is a deep, semicircular notch, the scapular notch, formed partly by the base of the coracoid process. This notch is converted into a foramen by the superior transverse scapular ligament, and serves for the passage of the suprascapular nerve; sometimes the ligament is ossified.
    - The adjacent part of the superior border affords attachment to the omohyoideus.
- The axillary border (or "lateral border") is the thickest of the three. It begins above at the lower margin of the glenoid cavity, and inclines obliquely downward and backward to the inferior angle. It is referred to as the caudal border in animals.
- It begins above at the lower margin of the glenoid cavity, and inclines obliquely downward and backward to the inferior angle. Immediately below the glenoid cavity is a rough impression, the infraglenoid tuberosity, about 2.5 cm (1 in). in length, which gives origin to the long head of the triceps brachii; in front of this is a longitudinal groove, which extends as far as the lower third of this border, and affords origin to part of the subscapularis. The inferior third is thin and sharp, and serves for the attachment of a few fibers of the teres major behind, and of the subscapularis in front.
- The medial border (also called the vertebral border or medial margin) is the longest of the three borders, and extends from the superior angle to the inferior angle. In animals it is referred to as the dorsal border. Four muscles attach

to the medial border. Serratus anterior has a long attachment on the anterior lip. Three muscles insert along the posterior lip, the levator scapulae (uppermost), rhomboid minor (middle), and to the rhomboid major (lower middle).

Development

The scapula is ossified from 7 or more centers: one for the body, two for the coracoid process, two for the acromion, one for the vertebral border, and one for the inferior angle. Ossification of the body begins about the second month of fetal life, by an irregular quadrilateral plate of bone forming, immediately behind the glenoid cavity. This plate extends to form the chief part of the bone, the scapular spine growing up from its dorsal surface about the third month. Ossification starts as membranous ossification before birth. After birth, the cartilaginous components would undergo endochondral ossification. The larger part of the scapula undergoes membranous ossification. Some of the outer parts of the scapula are cartilaginous at birth, and would therefore undergo endochondral ossification.

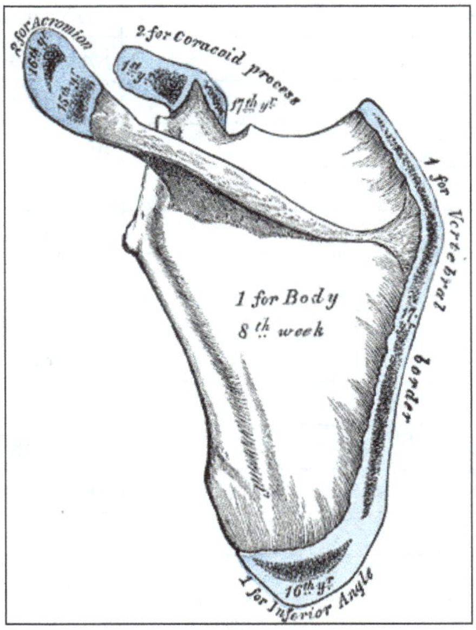

Plan of ossification of the scapula. From seven centers.

At birth, a large part of the scapula is osseous, but the glenoid cavity, the coracoid process, the acromion, the vertebral border and the inferior angle are cartilaginous. From the 15th to the 18th month after birth, ossification takes place in the middle of the coracoid process, which as a rule becomes joined with the rest of the bone about the 15th year.

Between the 14th and 20th years, the remaining parts ossify in quick succession, and usually in the following order: first, in the root of the coracoid process, in the form of a broad scale; secondly, near the base of the acromion; thirdly, in the inferior angle and contiguous part of the vertebral border; fourthly, near the outer end of the acromion;

fifthly, in the vertebral border. The base of the acromion is formed by an extension from the spine; the two nuclei of the acromion unite, and then join with the extension from the spine. The upper third of the glenoid cavity is ossified from a separate center (sub coracoid), which appears between the 10th and 11th years and joins between the 16th and the 18th years. Further, an epiphysial plate appears for the lower part of the glenoid cavity, and the tip of the coracoid process frequently has a separate nucleus. These various epiphyses are joined to the bone by the 25th year.

Failure of bony union between the acromion and spine sometimes occurs, the junction being effected by fibrous tissue, or by an imperfect articulation; in some cases of supposed fracture of the acromion with ligamentous union, it is probable that the detached segment was never united to the rest of the bone.

Function

The following muscles attach to the scapula:

Muscle	Direction	Region
Pectoralis Minor	insertion	coracoid process
Coracobrachialis	origin	coracoid process
Serratus Anterior	insertion	medial border
Triceps Brachii (long head)	origin	infraglenoid tubercle
Biceps Brachii (short head)	origin	coracoid process
Biceps Brachii (long head)	origin	supraglenoid tubercle
Subscapularis	origin	subscapular fossa
Rhomboid Major	insertion	medial border
Rhomboid Minor	insertion	medial border
Levator Scapulae	insertion	medial border
Trapezius	insertion	spine of scapula
Deltoid	origin	spine of scapula
Supraspinatus	origin	supraspinous fossa
Infraspinatus	origin	infraspinous fossa
Teres Minor	origin	lateral border
Teres Major	origin	lateral border
Latissimus Dorsi (a few fibers, attachment may be absent)	origin	inferior angle
Omohyoid	origin	superior border

Movements

Movements of the scapula are brought about by the scapular muscles. The scapula can perform six actions:

- Elevation: Upper trapezius and levator scapulae.

- Depression: Lower trapezius.
- Retraction (adduction): Rhomboids and middle trapezius.
- Protraction (abduction): Serratus anterior.
- Upward rotation: Upper and middle trapezius.
- Downward rotation: Rhomboids.

Clinical Significance

Scapular Fractures

Because of its sturdy structure and protected location, fractures of the scapula are uncommon. When they do occur, they are an indication that severe chest trauma has occurred. Scapular fractures involving the neck of the scapula have two patterns. One (rare) type of fracture is through the anatomical neck of the scapula. The other more common type of fracture is through the surgical neck of the scapula. The surgical neck exits medial to the coracoid process.

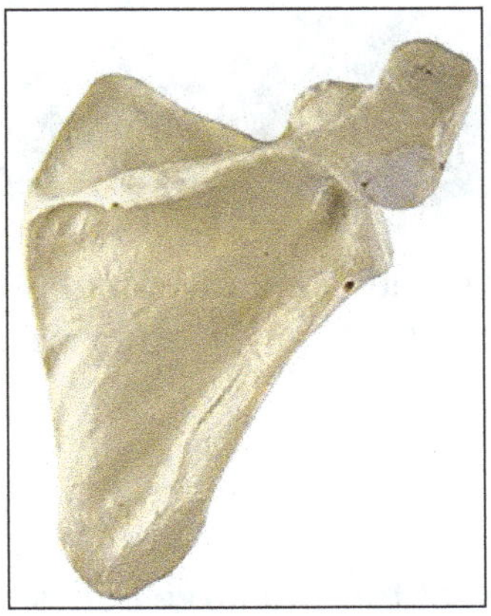

Left scapula, anterior surface: Anatomic neck: red, Surgical neck: purple.

An abnormally protruding inferior angle of the scapula is known as a winged scapula and can be caused by paralysis of the serratus anterior muscle. In this condition the sides of the scapula nearest the spine are positioned outward and backward. The appearance of the upper back is said to be wing-like. In addition, any condition causing weakness of the serratus anterior muscle may cause scapular "winging".

Impingement Syndrome

The scapula plays an important role in shoulder impingement syndrome.

Abnormal scapular function is called scapular dyskinesis. One action the scapula performs during a throwing or serving motion is elevation of the acromion process in order to avoid impingement of the rotator cuff tendons. If the scapula fails to properly elevate the acromion, impingement may occur during the cocking and acceleration phase of an overhead activity. The two muscles most commonly inhibited during this first part of an overhead motion are the serratus anterior and the lower trapezius. These two muscles act as a force couple within the glenohumeral joint to properly elevate the acromion process, and if a muscle imbalance exists, shoulder impingement may develop.

Humerus

The humerus is the longest and largest bone of the upper limb. It consists of a proximal end, a shaft and a distal end, all which contain important anatomical landmarks. The humerus articulates with the scapula proximally at the glenohumeral joint so it participates in the movements of the shoulder. Also, the humerus has distal articulations with the radius and ulna at the elbow joint.

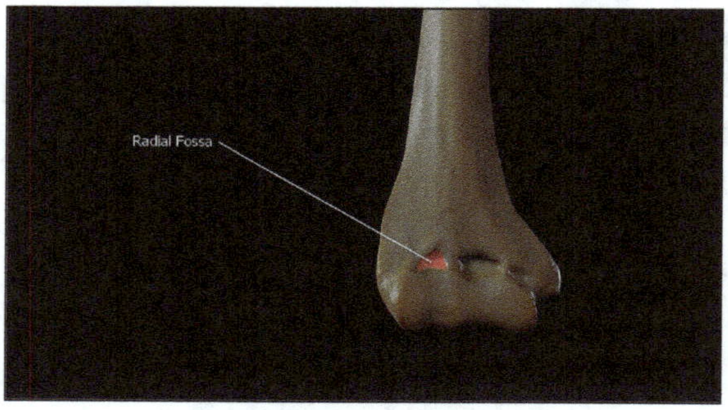

The nature of the elbow joint enables the movements that are limited to the arm and forearm, and cannot be performed within the other parts of the body, such as supination and pronation.

Radius

The radius or radial bone is one of the two large bones of the forearm, the other being the ulna. It extends from the lateral side of the elbow to the thumb side of the wrist and runs parallel to the ulna. The ulna is shorter and smaller than the radius. It is a long bone, prism-shaped and slightly curved longitudinally.

The radius is part of two joints: the elbow and the wrist. At the elbow, it joins with the capitulum of the humerus, and in a separate region, with the ulna at the radial notch. At the wrist, the radius forms a joint with the ulna bone.

The corresponding bone in the lower leg is the fibula.

Structure

The long narrow medullary cavity is enclosed in a strong wall of compact bone. It is thickest along the interosseous border and thinnest at the extremities, same over the cup-shaped articular surface (fovea) of the head.

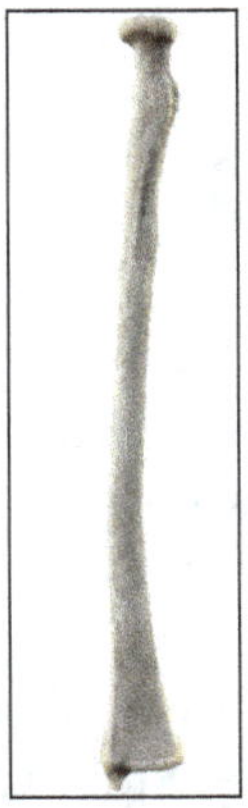

Full Anterior View of Right Radius.

The trabeculae of the spongy tissue are somewhat arched at the upper end and pass upward from the compact layer of the shaft to the fovea capituli (the humerus's cup-shaped articulatory notch); they are crossed by others parallel to the surface of the fovea. The arrangement at the lower end is somewhat similar. It is missing in radial aplasia.

The radius has a body and two extremities. The upper extremity of the radius consists of a somewhat cylindrical head articulating with the ulna and the humerus, a neck, and a radial tuberosity. The body of the radius is self-explanatory, and the lower extremity of the radius is roughly quadrilateral in shape, with articular surfaces for the ulna, scaphoid and lunate bones. The distal end of the radius forms two palpable points, radially the styloid process and Lister's tubercle on the ulnar side. Along with the proximal and distal radioulnar articulations, an interosseous membrane originates medially along the length of the body of the radius to attach the radius to the ulna.

Near the Wrist

The distal end of the radius is large and of quadrilateral form.

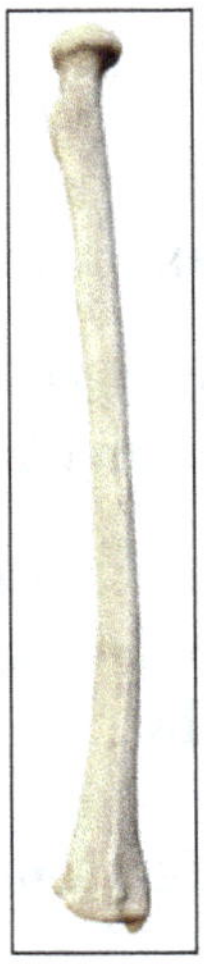

Full Posterior View of Right Radius.

Joint Surfaces

It is provided with two articular surfaces – one below, for the carpus, and another at the medial side, for the ulna.

- The carpal articular surface is triangular, concave, smooth, and divided by a slight antero-posterior ridge into two parts. Of these, the lateral, triangular, articulates with the scaphoid bone; the medial, quadrilateral, with the lunate bone.
- The articular surface for the ulna is called the ulnar notch (sigmoid cavity) of the radius; it is narrow, concave, smooth, and articulates with the head of the ulna.

These two articular surfaces are separated by a prominent ridge, to which the base of the triangular articular disk is attached; this disk separates the wrist-joint from the distal radioulnar articulation.

Other Surfaces

This end of the bone has three non-articular surfaces – volar, dorsal, and lateral.

- The volar surface, rough and irregular, affords attachment to the volar radiocarpal ligament.
- The dorsal surface is convex, affords attachment to the dorsal radiocarpal ligament, and is marked by three grooves. Enumerated from the lateral side:
    - The first groove is broad, but shallow, and subdivided into two by a slight ridge: the lateral of these two, transmits the tendon of the extensor carpi radialis longus muscle; the medial, the tendon of the extensor carpi radialis brevis muscle.
    - The second is deep but narrow, and bounded laterally by a sharply defined ridge; it is directed obliquely from above downward and lateralward, and transmits the tendon of the extensor pollicis longus muscle.
    - The third is broad, for the passage of the tendons of the extensor indicis proprius and extensor digitorum communis.
- The lateral surface is prolonged obliquely downward into a strong, conical projection, the styloid process, which gives attachment by its base to the tendon of the brachioradialis, and by its apex to the radial collateral ligament of wrist joint. The lateral surface of this process is marked by a flat groove, for the tendons of the abductor pollicis longus muscle and extensor pollicis brevis muscle.

Body

The body of the radius (or shaft of radius) is prismoid in form, narrower above than below, and slightly curved, so as to be convex lateralward. It presents three borders and three surfaces.

Borders

- The volar border (margo volaris; anterior border; palmar;) extends from the lower part of the tuberosity above to the anterior part of the base of the styloid process below, and separates the volar from the lateral surface. Its upper third is prominent, and from its oblique direction has received the name of the oblique line of the radius; it gives origin to the flexor digitorum superficialis muscle (also flexor digitorum sublimis) and flexor pollicis longus muscle; the surface above the line gives insertion to part of the supinator muscle. The middle third of the volar border is indistinct and rounded. The lower fourth is prominent, and gives insertion to the pronator quadratus muscle, and attachment to the dorsal carpal ligament; it ends in a small tubercle, into which the tendon of the brachioradialis muscle is inserted.
- The dorsal border (margo dorsalis; posterior border) begins above at the back of the neck, and ends below at the posterior part of the base of the styloid process; it separates the posterior from the lateral surface. is indistinct above and below, but well-marked in the middle third of the bone.
- The interosseous border (internal border; crista interossea; interosseous crest;) begins above, at the back part of the tuberosity, and its upper part is rounded and indistinct; it becomes sharp and prominent as it descends, and at its lower part divides into two ridges which are continued to the anterior and posterior margins of the ulnar notch. To the posterior of the two ridges the lower part of the interosseous membrane is attached, while the triangular surface between the ridges gives insertion to part of the pronator quadratus muscle. This crest separates the volar from the dorsal surface, and gives attachment to the interosseous membrane. The connection between the two bones is actually a joint referred to as a syndesmosis joint.

Surfaces

The volar surface (facies volaris; anterior surface) is concave in its upper three-fourths, and gives origin to the flexor pollicis longus muscle; it is broad and flat in its lower fourth, and affords insertion to the Pronator quadratus. A prominent ridge limits the insertion of the Pronator quadratus below, and between this and the inferior border is a triangular rough surface for the attachment of the volar radiocarpal ligament. At the junction of the upper and middle thirds of the volar surface is the nutrient foramen, which is directed obliquely upward.

The dorsal surface (facies dorsalis; posterior surface) is convex, and smooth in the upper third of its extent, and covered by the Supinator. Its middle third is broad, slightly concave, and gives origin to the Abductor pollicis longus above, and the extensor pollicis brevis muscle below. Its lower third is broad, convex, and covered by the tendons of the muscles which subsequently run in the grooves on the lower end of the bone.

The lateral surface (facies lateralis; external surface) is convex throughout its entire extent and is known as the convexity of the radius, curving outwards to be convex at the side. Its upper third gives insertion to the supinator muscle. About its center is a rough ridge, for the insertion of the pronator teres muscle. Its lower part is narrow, and covered by the tendons of the abductor pollicis longus muscle and extensor pollicis brevis muscle.

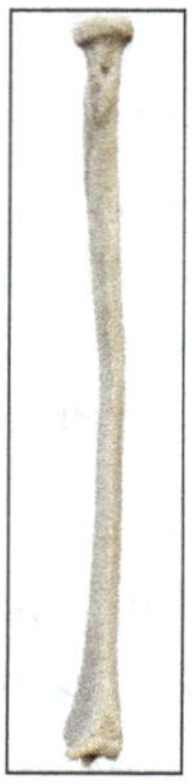

Full Lateral View of Right Radius.

Near the Elbow

The upper extremity of the radius (or proximal extremity) presents a head, neck, and tuberosity:

- The radial head has a cylindrical form, and on its upper surface is a shallow cup or fovea for articulation with the capitulum (or capitellum) of the humerus. The circumference of the head is smooth; it is broad medially where it articulates with the radial notch of the ulna, narrow in the rest of its extent, which is embraced by the annular ligament. The deepest point in the fovea is not axi-symmetric with the long axis of the radius, creating a cam effect during pronation and supination.
- The head is supported on a round, smooth, and constricted portion called the *neck*, on the back of which is a slight ridge for the insertion of part of the supinator muscle.
- Beneath the neck, on the medial side, is an eminence, the radial tuberosity; its surface is divided into a posterior, rough portion, for the insertion of the tendon of the biceps brachii muscle, and an anterior, smooth portion, on which a bursa is interposed between the tendon and the bone.

Development

The radius is ossified from *three* centers: one for the body, and one for each extremity. That for the body makes its appearance near the center of the bone, during the eighth week of fetal life. Ossification commences in the lower end between 9 and 26 months of age. The ossification center for the upper end appears by the fifth year. The upper epiphysis fuses with the body at the age of seventeen or eighteen years, the lower about the age of twenty.

An additional center sometimes found in the radial tuberosity, appears about the fourteenth or fifteenth year.

Function

Muscle Attachments

The biceps muscle inserts on the radial tuberosity of the upper extremity of the bone. The upper third of the body of the bone attaches to the supinator, the flexor digitorum superficialis, and the flexor pollicis longus muscles. The middle third of the body attaches to the extensor ossis metacarpi pollicis, extensor primi internodii pollicis, and the pronator teres muscles. The lower quarter of the body attaches to the pronator quadratus muscle and the tendon of the supinator longus.

Clinical Significance

Radial aplasia refers to the congenital absence or shortness of the radius.

Fracture

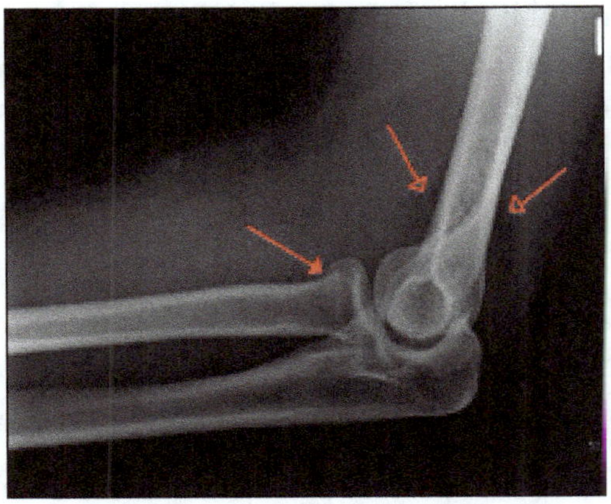

A subtle radial head fracture with associated positive sail sign.

Specific fracture types of the radius include:

- Proximal radius fracture: A fracture within the capsule of the elbow joint results in the fat pad sign or "sail sign" which is a displacement of the fat pad at the elbow.

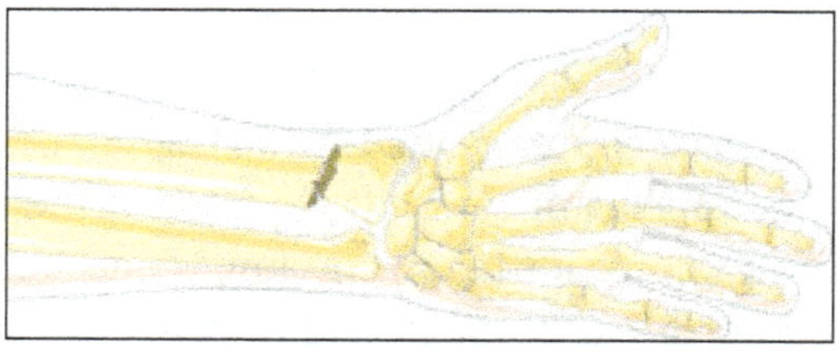

Illustration showing radius shaft fracture.

- Essex-Lopresti fracture: A fracture of the radial head with concomitant dislocation of the distal radio-ulnar joint with disruption of the interosseous membrane.
- Radial shaft fracture.
- Distal radius fracture:
    - Galeazzi fracture: A fracture of the radius with dislocation of the distal radioulnar joint.
    - Colles' fracture: A distal fracture of the radius with dorsal (posterior) displacement of the wrist and hand.
    - Smith's fracture: A distal fracture of the radius with volar (ventral) displacement of the wrist and hand.
    - Barton's fracture: An intra-articular fracture of the distal radius with dislocation of the radiocarpal joint.

Ulna

The ulna is a long bone found in the forearm that stretches from the elbow to the smallest finger, and when in anatomical position, is found on the medial side of the forearm. It runs parallel to the radius, the other long bone in the forearm, and is the larger and longer of the two.

Structure

The ulna is a long bone found in the forearm that stretches from the elbow to the smallest finger, and when in anatomical position, is found on the medial side of the forearm. It is broader close to the elbow, and narrows as it approaches the wrist.

Close to the elbow, the ulna has a bony process, the olecranon process, a hook-like structure that fits into the olecranon fossa of the humerus. This prevents hyperextension and forms a hinge joint with the trochlea of the humerus. There is also a radial notch for the head of the radius, and the ulnar tuberosity to which muscles attach.

Close to the wrist, the ulna has a styloid process.

Lower End of Ulna

Near the Elbow

Near the elbow, the ulna has two curved processes, the olecranon and the coronoid process; and two concave, articular cavities, the semilunar and radial notches.

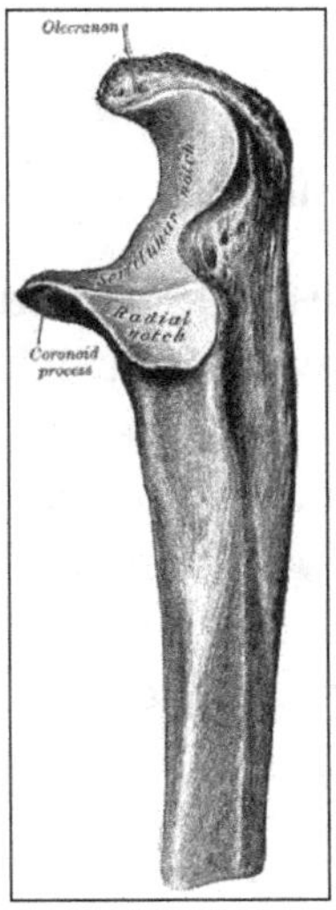

Upper extremity of left ulna - Lateral aspect.

The olecranon is a large, thick, curved eminence, situated at the upper and back part of the ulna. It is bent forward at the summit so as to present a prominent lip which is received into the olecranon fossa of the humerus in extension of the forearm. Its base is contracted where it joins the body and the narrowest part of the upper end of the ulna. Its posterior surface, directed backward, is triangular, smooth, subcutaneous, and covered by a bursa. Its superior surface is of quadrilateral form, marked behind by a rough impression for the insertion of the triceps brachii; and in front, near the margin, by a slight transverse groove for the attachment of part of the posterior ligament of the elbow joint. Its anterior surface is smooth, concave, and forms the upper part of the semilunar notch. Its borders present continuations of the groove on the margin of the superior surface; they serve for the attachment of ligaments: the back part of the ulnar collateral ligament medially, and the posterior ligament laterally. From the medial border a part of the flexor carpi ulnaris arises; while to the lateral border the anconeus is attached.

The coronoid process is a triangular eminence projecting forward from the upper and front part of the ulna. Its base is continuous with the body of the bone, and of considerable strength. Its apex is pointed, slightly curved upward, and in flexion of the forearm is received into the coronoid fossa of the humerus. Its upper surface is smooth, concave, and forms the lower part of the semilunar notch. Its antero-inferior surface is concave, and marked by a rough impression for the insertion of the brachialis. At the junction of this surface with the front of the body is a rough eminence, the tuberosity of the ulna, which gives insertion to a part of the brachialis; to the lateral border of this tuberosity the oblique cord is attached. Its lateral surface presents a narrow, oblong, articular depression, the radial notch. Its medial surface, by its prominent, free margin, serves for the attachment of part of the ulnar collateral ligament. At the front part of this surface is a small rounded eminence for the origin of one head of the flexor digitorum superficialis; behind the eminence is a depression for part of the origin of the flexor digitorum profundus; descending from the eminence is a ridge which gives origin to one head of the pronator teres. Frequently, the flexor pollicis longus arises from the lower part of the coronoid process by a rounded bundle of muscular fibers.

The semilunar notch is a large depression, formed by the olecranon and the coronoid process, and serving as articulation with the trochlea of the humerus. About the middle of either side of this notch is an indentation, which contracts it somewhat, and indicates the junction of the olecranon and the coronoid process. The notch is concave from above downward, and divided into a medial and a lateral portion by a smooth ridge running from the summit of the olecranon to the tip of the coronoid process. The medial portion is the larger, and is slightly concave transversely; the lateral is convex above, slightly concave below.

The radial notch is a narrow, oblong, articular depression on the lateral side of the coronoid process; it receives the circumferential articular surface of the head of the radius. It is concave from before backward, and its prominent extremities serve for the attachment of the annular ligament.

Body

The body of the ulna at its upper part is prismatic in form, and curved so as to be convex behind and lateralward; its central part is straight; its lower part is rounded, smooth, and bent a little lateralward. It tapers gradually from above downward, and has three borders and three surfaces.

Borders

- The volar border (margo volaris; anterior border) begins above at the prominent medial angle of the coronoid process, and ends below in front of the styloid process. Its upper part, well-defined, and its middle portion, smooth and rounded, give origin to the flexor digitorum profundus; its lower fourth serves for the origin of the pronator quadratus. This border separates the volar from the medial surface.
- The dorsal border (margo dorsalis; posterior border) begins above at the apex of the triangular subcutaneous surface at the back part of the olecranon, and ends below at the back of the styloid process; it is well-marked in the upper three-fourths, and gives attachment to an aponeurosis which affords a common origin

to the flexor carpi ulnaris, the extensor carpi ulnaris, and the flexor digitorum profundus; its lower fourth is smooth and rounded. This border separates the medial from the dorsal surface.

- The interosseous crest (crista interossea; external or interosseous border) begins above by the union of two lines, which converge from the extremities of the radial notch and enclose between them a triangular space for the origin of part of the Supinator; it ends below at the head of the ulna. Its upper part is sharp, its lower fourth smooth and rounded. This crest gives attachment to the interosseous membrane, and separates the volar from the dorsal surface.

Surfaces

- The volar surface (facies volaris; anterior surface), much broader above than below, is concave in its upper three-fourths, and gives origin to the flexor digitorum profundus; its lower fourth, also concave, is covered by the pronator quadratus. The lower fourth is separated from the remaining portion by a ridge, directed obliquely downward and medialward, which marks the extent of origin of the pronator quadratus. At the junction of the upper with the middle third of the bone is the nutrient canal, directed obliquely upward.
- The dorsal surface (facies dorsalis; posterior surface) directed backward and lateralward, is broad and concave above; convex and somewhat narrower in the middle; narrow, smooth, and rounded below. On its upper part is an oblique ridge, which runs from the dorsal end of the radial notch, downward to the dorsal border; the triangular surface above this ridge receives the insertion of the Anconæus, while the upper part of the ridge affords attachment to the supinator. Below this the surface is subdivided by a longitudinal ridge, sometimes called the perpendicular line, into two parts: the medial part is smooth, and covered by the extensor carpi ulnaris; the lateral portion, wider and rougher, gives origin from above downward to the Supinator, the abductor pollicis longus, the extensor pollicis longus, and the extensor indicis proprius.
- The medial surface (facies medialis; internal surface) is broad and concave above, narrow and convex below. Its upper three-fourths give origin to the Flexor digitorum profundus; its lower fourth is subcutaneous.

Near the Wrist

The head of ulna presents an articular surface, part of which, of an oval or semilunar form, is directed downward, and articulates with the upper surface of the triangular articular disc which separates it from the wrist-joint; the remaining portion, directed lateralward, is narrow, convex, and received into the ulnar notch of the radius.

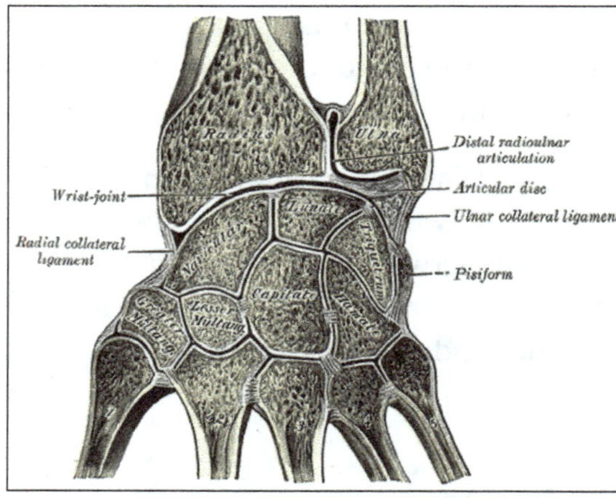

Vertical section through the articulations at the wrist, showing the synovial cavities.

Near the wrist, the ulnar, with two eminences; the lateral and larger is a rounded, articular eminence, termed the head of the ulna; the medial, narrower and more projecting, is a non-articular eminence, the styloid process.

- The head presents an articular surface, part of which, of an oval or semilunar form, is directed downward, and articulates with the upper surface of the triangular articular disk which separates it from the wrist-joint; the remaining portion, directed lateralward, is narrow, convex, and received into the ulnar notch of the radius.
- The styloid process projects from the medial and back part of the bone; it descends a little lower than the head, and its rounded end affords attachment to the ulnar collateral ligament of the wrist-joint.

The head is separated from the styloid process by a depression for the attachment of the apex of the triangular articular disk, and behind, by a shallow groove for the tendon of the extensor carpi ulnaris.

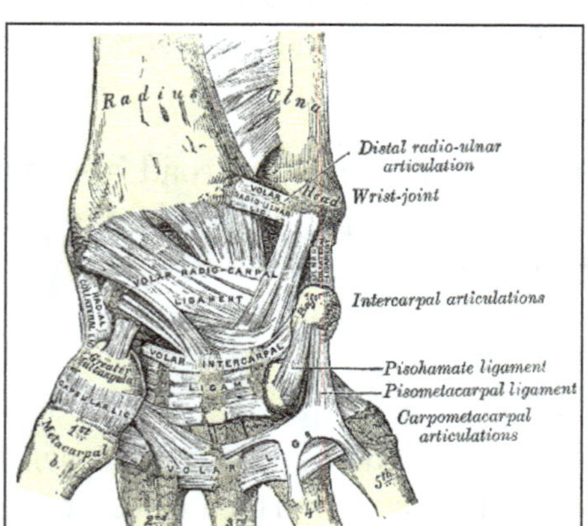

Ligaments of wrist. Anterior view.

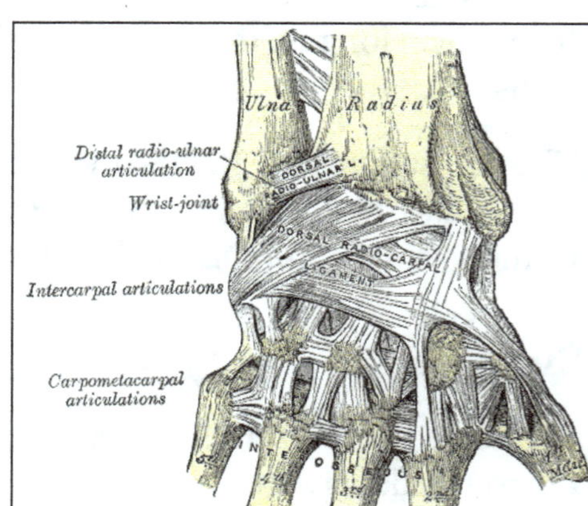

Ligaments of wrist. Posterior view.

Microanatomy

The ulna is a long bone. The long, narrow medullary cavity of the ulna is enclosed in a strong wall of cortical tissue which is thickest along the interosseous border and dorsal surface. At the extremities the compact layer thins. The compact layer is continued onto the back of the olecranon as a plate of close spongy bone with lamellæ parallel. From the inner surface of this plate and the compact layer below it trabeculæ arch forward toward the olecranon and coronoid and cross other trabeculæ, passing backward over the medullary cavity from the upper part of the shaft below the coronoid. Below the coronoid process there is a small area of compact bone from which trabeculæ curve upward to end obliquely to the surface of the semilunar notch which is coated with a thin layer of compact bone. The trabeculæ at the lower end have a more longitudinal direction.

Development

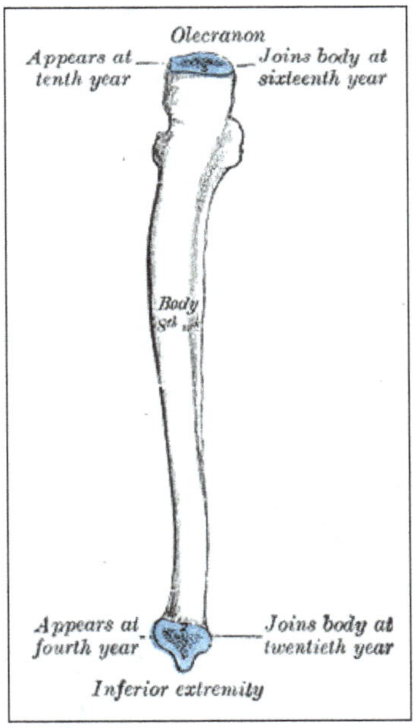

The ulna is formed into bone from three centres.

The ulna is ossified from three centers: one each for the body, the wrist end, and the elbow end, near the top of the olecranon. Ossification begins near the middle of the body of the ulna, about the eighth week of fetal life, and soon extends through the greater part of the bone.

At birth, the ends are cartilaginous. About the fourth year or so, a center appears in the middle of the head, and soon extends into the ulnar styloid process. About the tenth year, a center appears in the olecranon near its extremity, the chief part of this process being formed by an upward extension of the body. The upper epiphysis joins the body about the sixteenth, the lower about the twentieth year.

Function

Joints

The ulna forms part of the wrist joint and elbow joints. Specifically, the ulna joins (articulates) with:

- Trochlea of the humerus, at the right side elbow as a hinge joint with semilunar trochlear notch of the ulna.
- The radius, near the elbow as a pivot joint, this allows the radius to cross over the ulna in pronation.
- The distal radius, where it fits into the ulnar notch.
- The radius along its length via the interosseous membrane that forms a syndesmosis joint.

Muscle Attachments

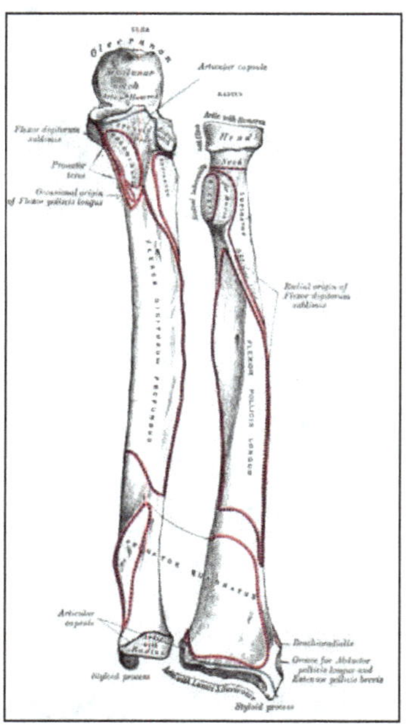

Bones of left forearm. Anterior aspect.

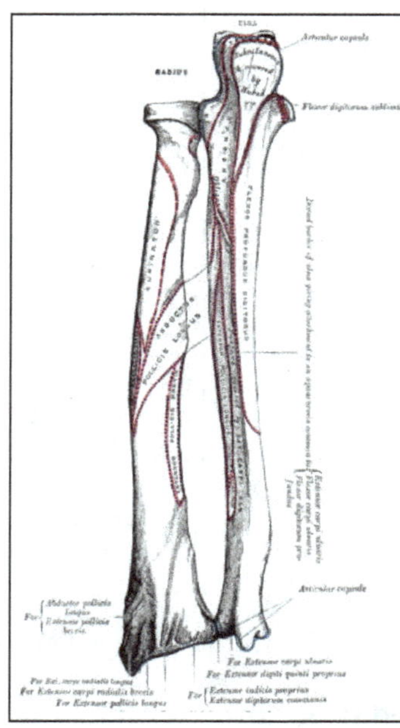

The radius and ulna of the left forearm, posterior surface.

Muscle	Direction	Attachment
Triceps brachii muscle	Insertion	posterior part of superior surface of Olecranon process (via common tendon)
Anconeus muscle	Insertion	olecranon process (lateral aspect)
Brachialis muscle	Insertion	anterior surface of the coronoid process of the ulna

Pronator teres muscle	Origin	medial surface on middle portion of coronoid process (also shares origin with medial epicondyle of the humerus)
Flexor carpi ulnaris muscle	Origin	olecranon process and posterior surface of ulna (also shares origin with medial epicondyle of the humerus)
Flexor digitorum superficialis muscle	Origin	coronoid process (also shares origin with medial epicondyle of the humerus and shaft of the radius)
Flexor digitorum profundus muscle	Origin	anteromedial surface of ulna (also shares origin with the interosseous membrane)
Pronator quadratus muscle	Origin	distal portion of anterior ulnar shaft
Extensor carpi ulnaris muscle	Origin	posterior border of ulna (also shares origin with lateral epicondyle of the humerus)
Supinator muscle	Origin	proximal ulna (also shares origin with lateral epicondyle of the humerus)
Abductor pollicis longus muscle	Origin	posterior surface of ulna (also shares origin with the posterior surface of the radius bone)
Extensor pollicis longus muscle	Origin	dorsal shaft of ulna (also shares origin with the dorsal shaft of the radius and the interosseous membrane)
Extensor indicis muscle	Origin	posterior surface of distal ulna (also shares origin with the interosseous membrane)

Carpal Bones

The carpal bones are the eight small bones that make up the wrist (or carpus) that connects the hand to the forearm. In tetrapods, the carpus is the sole cluster of bones in the wrist between the radius and ulna and the metacarpus. The bones of the carpus do not belong to individual fingers (or toes in quadrupeds), whereas those of the metacarpus do. The corresponding part of the foot is the tarsus. The carpal bones allow the wrist to move and rotate vertically.

The eight carpal bones may be conceptually organized as either two transverse rows, or three longitudinal columns.

When considered as paired rows, each row forms an arch which is convex proximally and concave distally. On the palmar side, the carpus is concave and forms the carpal tunnel, which is covered by the flexor retinaculum. The proximal row (comprising scaphoid, lunate, and triquetrum) articulates with the surfaces of the radius and distal carpal row, and thus constantly adapts to these mobile surfaces. Within the proximal row, each carpal bone has slight independent mobility. For example, the scaphoid contributes to midcarpal stability by articulating distally with the trapezium and the trapezoid. In contrast, the distal row is more rigid as its transverse arch moves with the metacarpals.

Biomechanically and clinically, the carpal bones are better conceptualized as three longitudinal columns:

- Radial scaphoid column: Scaphoid, trapezium, and trapezoid.

- Lunate column: Lunate and capitate.
- Ulnar triquetral column: Triquetrum and hamate.

In this context the pisiform is regarded as a sesamoid bone embedded in the tendon of the flexor carpi ulnaris. The ulnar column leaves a gap between the ulna and the triquetrum, and therefore, only the radial or scaphoid and central or capitate columns articulate with the radius. The wrist is more stable in flexion than in extension more because of the strength of various capsules and ligaments than the interlocking parts of the skeleton.

Almost all carpals (except the pisiform) have six surfaces. Of these the palmar or anterior and the dorsal or posterior surfaces are rough, for ligamentous attachment; the dorsal surfaces being the broader, except in the lunate.

The superior or proximal, and inferior or distal surfaces are articular, the superior generally convex, the inferior concave; the medial and lateral surfaces are also articular where they are in contact with contiguous bones, otherwise they are rough and tuberculated.

The structure in all is similar: cancellous tissue enclosed in a layer of compact bone.

Bones

Table: Articulations of individual carpal bones.

Name	Proximal/radial articulations	Lateral/medial articulations	Distal/metacarpal articulations
Proximal row			
Scaphoid	radius	capitate, lunate	trapezium, trapezoid
Lunate	radius, articular disk	scaphoid, triquetral	capitate, hamate (sometimes)
Triquetrum	articular disk	lunate, pisiform	hamate
Pisiform		triquetral	
Distal row			
Trapezium	scaphoid	trapezoid	first and second metacarpal
Trapezoid	scaphoid	trapezium, capitate	second metacarpal
Capitate	scaphoid, lunate	trapezoid, hamate	third, partly second and fourth metacarpal
Hamate	triquetral, lunate	capitate	fourth and fifth

Accessory Bones

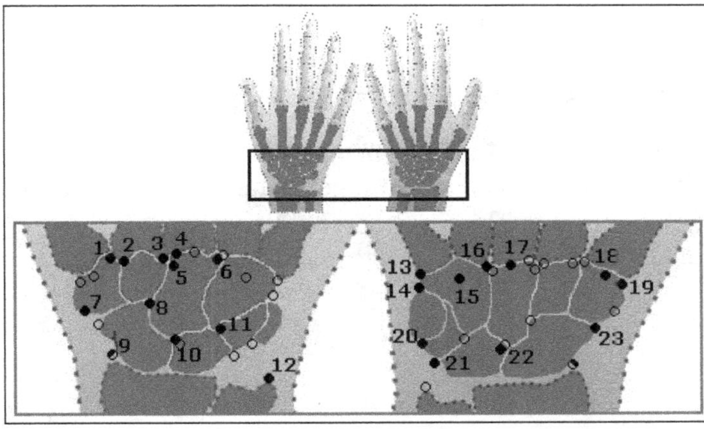

Location of the accessory ossicles of the carpals.

Occasionally accessory bones are found in the carpus, but of more than 20 such described bones, only four (the central, styloid, secondary trapezoid, and secondary pisiform bones) are considered to be proven accessory bones. Sometimes the scaphoid, triquetrum, and pisiform bones are divided into two.

Development

Table: Appearance of ossification centers of carpal bones.

Bone	Average	Variation	Variation
Capitate	2.5 months	1–6 months	1–5 months
Hamate	4-5.5 months	1–7 months	1–12 months
Triquetrum	2 years	5 months to 3 years	9 months to 4 years and 2 months
Lunate	5 years	2-5.5 years	18 months to 4 years and 3 months
Trapezium	6 years	4–8 years	
Trapezoid	6 years	4–8 years	
Scaphoid	6 years	4–7 years	
Pisiform	12 years	8–12 years	

The carpal bones are ossified endochondrally (from within the cartilage) and the ossific centers appear only after birth. The formation of these centers roughly follows a chronological spiral pattern starting in the capitate and hamate during the first year of life. The ulnar bones are then ossified before the radial bones, while the sesamoid pisiform arises in the tendon of the flexor carpi ulnaris after more than ten years. The commencement of ossification for each bone occurs over period like other bones. This is useful in forensic age estimation.

Function

Ligaments

There are four groups of ligaments in the region of the wrist:

- The ligaments of the wrist proper which unite the ulna and radius with the carpus: the ulnar and radial collateral ligaments; the palmar and dorsal radiocarpal ligaments; and the palmar ulnocarpal ligament.

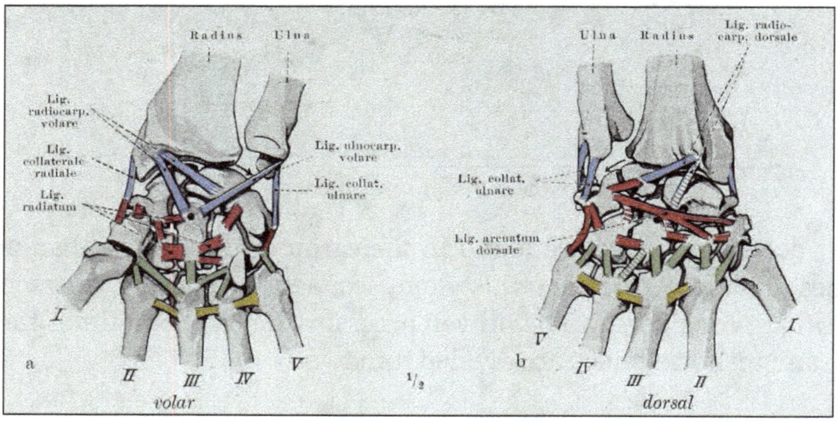

Four groups of ligaments in the region of the wrist (shown in four different colors).

- The ligaments of the intercarpal articulations which unite the carpal bones with one another: the radiate carpal ligament; the dorsal, palmar, and interosseous intercarpal ligaments; and the pisohamate ligament.
- The ligaments of the carpometacarpal articulations which unite the carpal bones with the metacarpal bones: the pisometacarpal ligament and the palmar and dorsal carpometacarpal ligaments.
- The ligaments of the intermetacarpal articulations which unite the metacarpal bones: the dorsal, interosseous, and palmar metacarpal ligaments.

Movements

The hand is said to be in straight position when the third finger runs over the capitate bone and is in a straight line with the forearm. This should not be confused with the midposition of the hand which corresponds to an ulnar deviation of 12 degrees. From the straight position two pairs of movements of the hand are possible: abduction (movement towards the radius, so called radial deviation or abduction) of 15 degrees and adduction (movement towards the ulna, so called ulnar deviation or adduction) of 40 degrees when the arm is in strict supination and slightly greater in strict pronation. Flexion (tilting towards the palm, so called palmar flexion) and extension (tilting towards the back of the hand, so called dorsiflexion) is possible with a total range of 170 degrees.

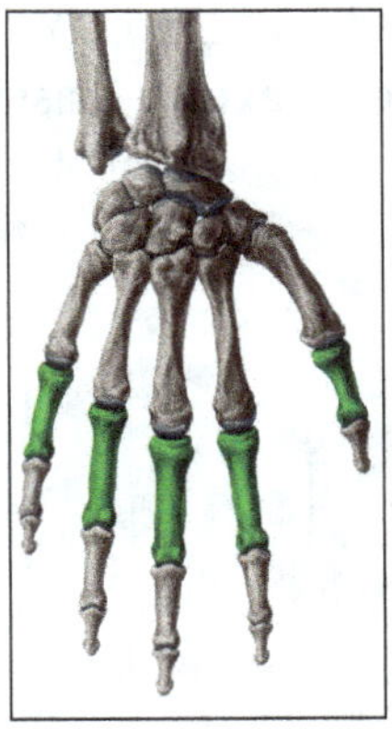

Proximal phalanges of the hand (dorsal view).

Radial Abduction/Ulnar Adduction

During radial abduction the scaphoid is tilted towards the palmar side which allows the trapezium and trapezoid to approach the radius. Because the trapezoid is rigidly attached to the second metacarpal bone to which also the flexor carpi radialis and extensor carpi radialis are attached, radial abduction effectively pulls this combined structure towards the radius. During radial abduction the pisiform traverses the greatest path of all carpal bones. Radial abduction is produced by (in order of importance) extensor carpi radialis longus, abductor pollicis longus, extensor pollicis longus, flexor carpi radialis, and flexor pollicis longus.

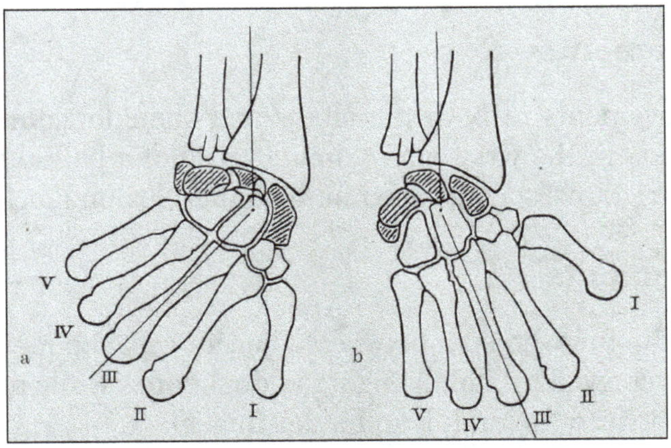

Left: Ulnar adduction. Right: Radial abduction.

Ulnar adduction causes a tilting or dorsal shifting of the proximal row of carpal bones. It is produced by extensor carpi ulnaris, flexor carpi ulnaris, extensor digitorum, and extensor digiti minimi. Both radial abduction and ulnar adduction occurs around a dorsopalmar axis running through the head of the capitate bone.

Palmar Flexion/Dorsiflexion

During palmar flexion the proximal carpal bones are displaced towards the *dorsal* side and towards the palmar side during dorsiflexion. While flexion and extension consist of movements around a pair of transverse axes – passing through the lunate bone for the proximal row and through the capitate bone for the distal row – palmar flexion occurs mainly in the radiocarpal joint and dorsiflexion in the midcarpal joint.

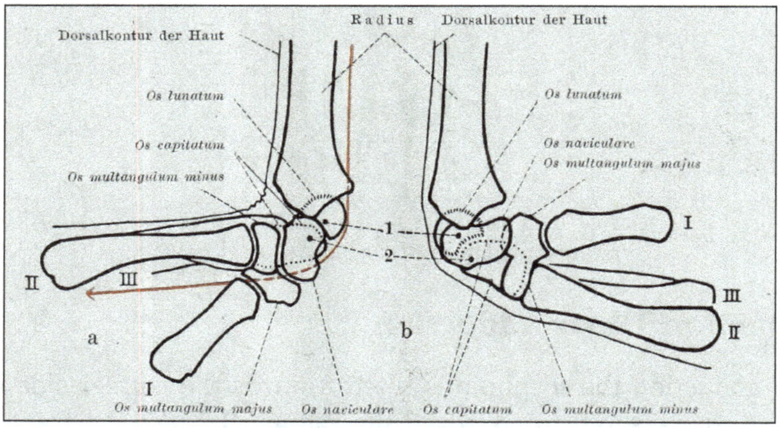

Left: Dorsiflexion. Right: Palmar flexion.

Dorsiflexion is produced by (in order of importance) extensor digitorum, extensor carpi radialis longus, extensor carpi radialis brevis, extensor indicis, extensor pollicis longus, and extensor digiti minimi. Palmar flexion is produced by (in order of importance) flexor digitorum superficialis, flexor digitorum profundus, flexor carpi ulnaris, flexor pollicis longus, flexor carpi radialis, and abductor pollicis longus.

Combined Movements

Combined with movements in both the elbow and shoulder joints, intermediate or combined movements in the wrist approximate those of a ball-and-socket joint with some necessary restrictions, such as maximum palmar flexion blocking abduction.

Accessory Movements

Anteroposterior gliding movements between adjacent carpal bones or along the midcarpal joint can be achieved by stabilizing individual bones while moving another (i.e. gripping the bone between the thumb and index finger).

Phalanges

The phalanges are the terminal bones of the hand. There are fourteen in total, with each finger having three; a proximal, middle and distal phalanx, with the exception of the thumb, which only has two; a proximal and a distal phalanx.

Proximal Phalanges

There are five proximal phalanges which correspond to each of the five fingers. Like the metacarpal bones, the proximal phalanges have a proximal base with a transverse oval articular facet which connects them to their subsequent metacarpal bones respectfully. Their shafts are long and are flat on their palmar aspect. They are convex both dorsally and transversally and also have sharpened medial and lateral borders in order for the fibrous tendon sheaths of the flexor muscles to attach. Distally, the trochlea or phalangeal head has an articular surface that links the proximal phalanges to the middle phalanges and in the case of the thumb, to its distal phalanx.

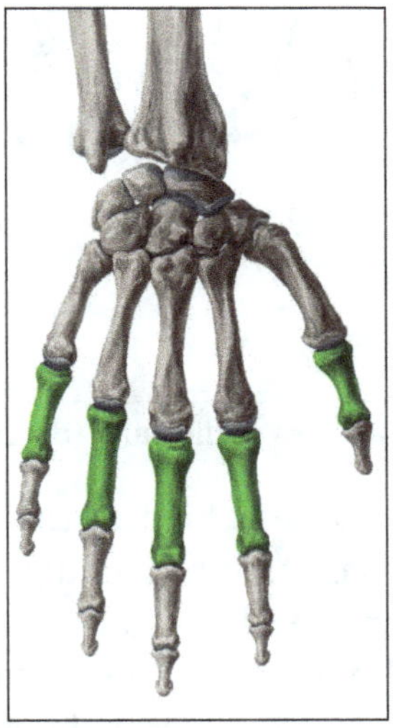

Proximal phalanges of the hand (dorsal view).

Middle Phalanges

There are four middle phalanges which correspond to the index finger, the middle finger, the ring finger and the little finger. They have a proximal base and a distal head which is separated by a much shorter shaft than that of the proximal phalanges. The base of a proximal phalanx has a medial and lateral facet that surrounds a midline groove that is smooth and is coherent with the distal head of the proximal phalanx. Distally, the heads of the middle phalanges articulate with the bases of the distal phalanges.

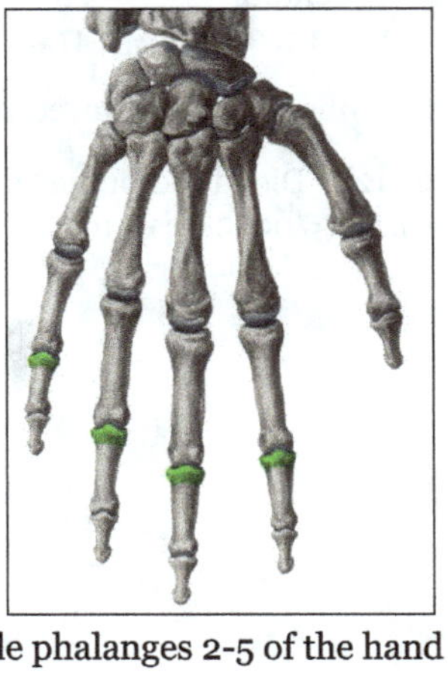

Bases of middle phalanges 2-5 of the hand (dorsal view).

Distal Phalanges

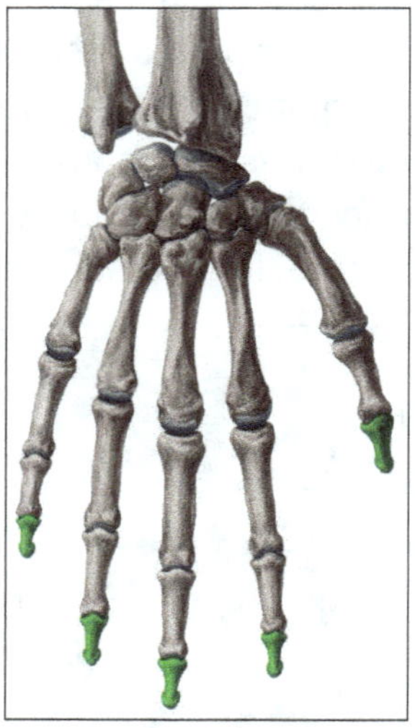

Distal phalanges of the hand (dorsal view).

There are five distal phalanges, including that of the thumb. Each of them is tapered distally and has a wider proximal base that articulates with the middle phalanges or the proximal phalanx in the case of the thumb. Distally, the surfaces of the distal phalanges are rough, especially on their palmar aspect where the tendon of the flexor digitorum profundus inserts. Dorsally, a rough spade-shaped plate that faces in the palmar direction can be seen and is known as the tuberosity of the distal phalanx.

References

- Anatomy-of-the-elbow, elbow, medical-services, orthopaedics, upper-body: baxterregional.org, Retrieved 24 February, 2020
- Moore, Keith; Anne Agur (2007). Essential Clinical Anatomy Third Edition. USA: Lippincott Williams & Wilkins. p. 446. ISBN 0-7817-6274-X
- 8-2-bones-of-the-upper-limb, anatomyandphysiology: opentextbc.ca, Retrieved 04 August, 2020
- Frich, Lars Henrik; Larsen, Morten Schultz (2017). "How to deal with a glenoid fracture". EFORT Open Reviews. 2 (5): 151–157. doi:10.1302/2058-5241.2.160082. ISSN 2396-7544
- The-humerus, anatomy, library: kenhub.com, Retrieved 16 June, 2020
- Thomas, B. P.; Sreekanth, R. (2012). "Distal radioulnar joint injuries". Indian Journal of Orthopaedics. 46 (5): 493–504. doi:10.4103/0019-5413.101031. PMC 3491781. PMID 23162140

5

Lower Limb Anatomy

The upper leg, the lower leg, and the foot together comprises the lower limb anatomy. The leg consists of two long bones, the tibia and fibula, and the sesamoid bone up. The patella serves as the knee-cap and the thigh consists of a single bone, the femur. This chapter sheds light on the subject of lower limb anatomy for a thorough understanding of the subject.

Thigh

There are three layers of gluteal muscles on the posterior hips, just like there are three layers of muscles in the abdominal trunk. The largest of them is the most superficial muscle, the gluteus maximus. Its origin is on the ilium of the coxal bone, and it inserts part-way down the shaft of the femur. It helps maintain erect posture, abducts the thigh, and rotates the thigh outward.

Below the gluteus maximus is the smaller gluteus medius. The gluteus medius muscle helps abducts the thigh along with the gluteus maximus, but can rotate the thigh inward where the gluteus maximus rotates the thigh outward.

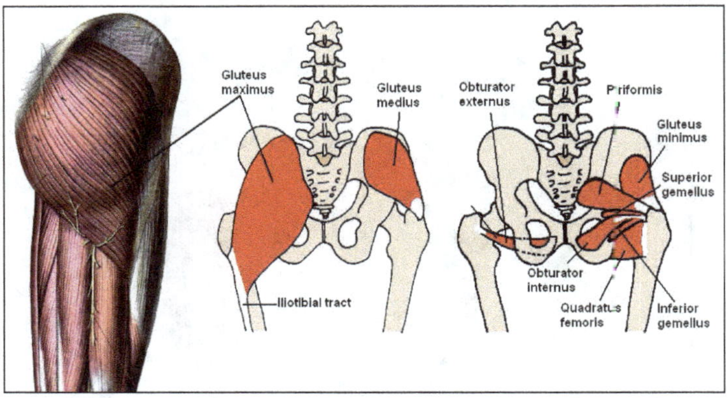

The three layers of gluteal muscles, gluteus maximus, gluteus medius, gluteus minimus.

The gluteus medius are several muscles, one of which is the gluteus minimus, the smallest of the gluteal muscles. It is a synergist for the gluteus medius.

Like the forearm, the upper leg, or thigh, has a dense arrangement of many muscles. On the anterior side, the most prominent of the muscles are the sartorius muscle and the four muscles that make up quadriceps muscle group (the "quads").

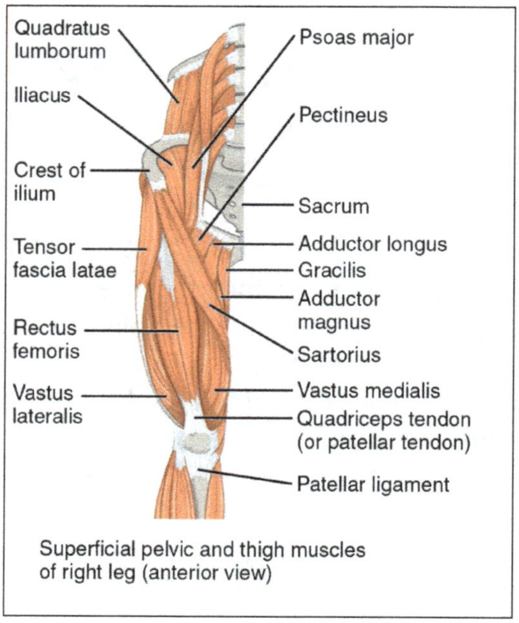

The superficial muscles of the thigh.

The quadriceps sounds like it should be just one muscle, akin to the triceps brachii, but it is a group of four muscles, three visible on the surface, and the fourth obscured. The three surface muscles of the quadriceps are the rectus femoris in the center, the vastus medialis on the medial side, and the vastus lateralis on the lateral side. Below the rectus femoris and largely hidden by it is the vastus intermedius. This muscle's position can be seen in figure. The four muscle of the quadriceps all extend the lower leg, and the rectus femoris additionally can flex the thigh at the hip.

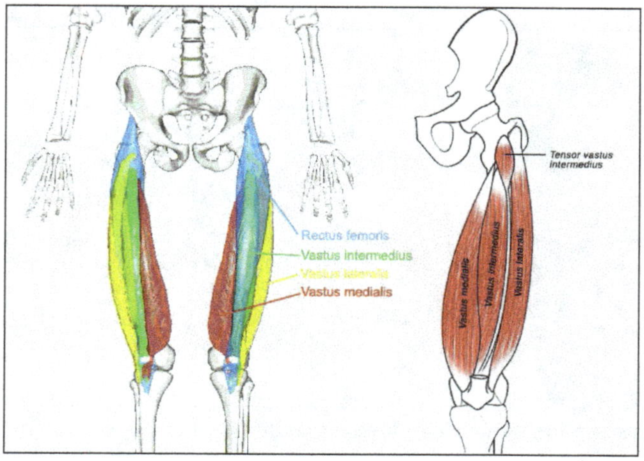

The quadriceps group of four muscles. The view on the left has the rectus femoris cut away to show the vastus intermedius which is below it.

The sartorius muscle is a distinctively long and thin muscle that crosses the thigh diagonally. It is visible in figure. Sartorius comes from the Latin for tailor, and this is sometimes called the tailor's muscle, although the reasons for the nickname are obscure. It may be because the shape of the muscle is thin and long, like a tailor's measuring tape; it may be because it is close to the inseam a tailor measures when tailoring pants, or it may be because it helps bring about the cross-legged position that tailors often adopt when working.

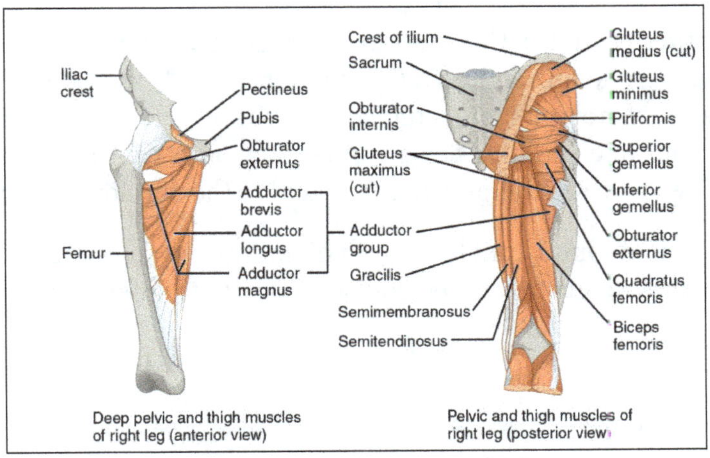

The muscles of the posterior thigh.

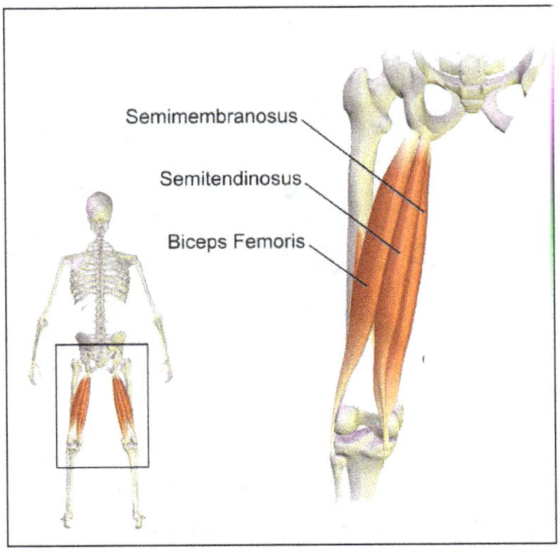

The hamstring group of muscles of the posterior thigh.

In the posterior thigh the bulk of the musculature is made up of three long muscles that are collectively called the hamstrings. The origin of this nickname is obscure, but it may have to do with the practice of butchers of hanging the thighs of butchered animals such as pig (the "hams") by the tendons of these three muscles. Move from the medial edge to the lateral edge of the posterior thigh, the hamstring muscles are the semimembranous muscle, the semitendinosus muscle, and the biceps femoris muscle. Notice the upper leg has a "biceps" muscle just like the upper arm does. This is why you have to indicate which biceps you are taking about when discussing one or other of these muscles. On the medial edge of the posterior thigh is the gracilis muscle. It is also visible on the medial edge of the thigh from the anterior.

Knee

In humans and other primates, the knee joins the thigh with the leg and consists of two joints: one between the femur and tibia (tibiofemoral joint), and one between the femur and patella (patellofemoral joint). It is the largest joint in the human body. The knee is a modified hinge joint, which permits flexion and extension as well as slight internal and external rotation. The knee is vulnerable to injury and to the development of osteoarthritis.

It is often termed a compound joint having tibiofemoral and patellofemoral components. (The fibular collateral ligament is often considered with tibiofemoral components).

Structure

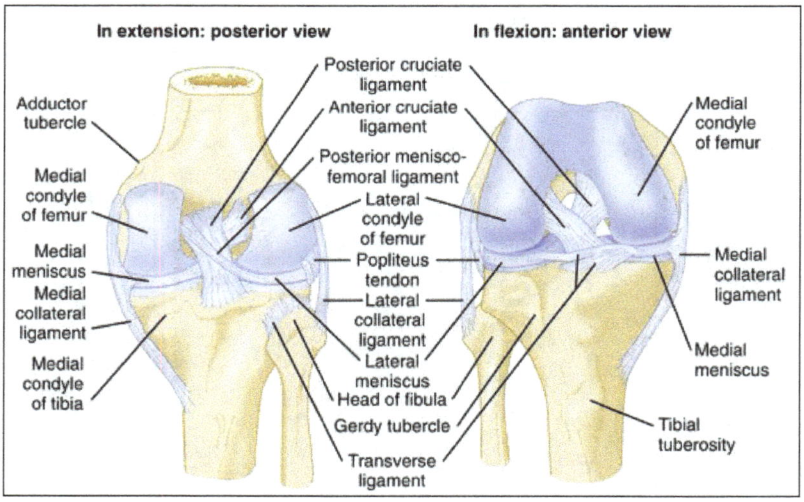

The knee is a modified hinge joint, a type of synovial joint, which is composed of three functional compartments: the patellofemoral articulation, consisting of the patella, or "kneecap", and the patellar groove on the front of the femur through which it slides; and the medial and lateral tibiofemoral articulations linking the femur, or thigh bone, with the tibia, the main bone of the lower leg. The joint is bathed in synovial fluid which is contained inside the synovial membrane called the joint capsule. The posterolateral corner of the knee is an area that has recently been the subject of renewed scrutiny and research.

The knee is the largest joint and one of the most important joints in the body. It plays an essential role in movement related to carrying the body weight in horizontal (running and walking) and vertical (jumping) directions. At birth, the kneecap is just formed from cartilage, and this will ossify (change to bone) between the ages of three and five years. Because it is the largest sesamoid bone in the human body, the ossification process takes significantly longer.

Articular Bodies

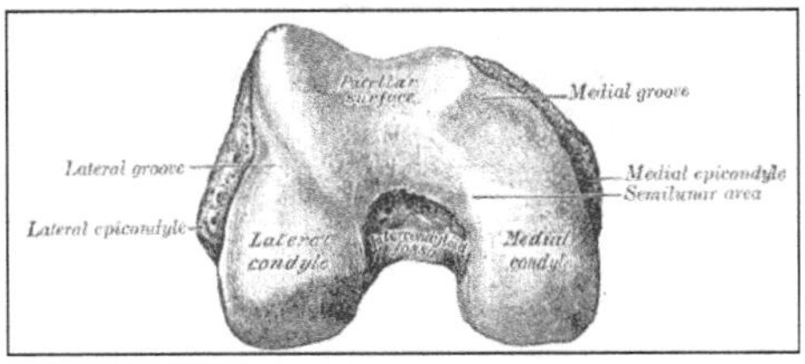

Articular surfaces of femur.

The main articular bodies of the femur are its lateral and medial condyles. These diverge slightly distally and posteriorly, with the lateral condyle being wider in front than at the back while the medial condyle is of more constant width. The radius of the condyles' curvature in the sagittal plane becomes smaller toward the back. This diminishing radius produces a series of involute midpoints (i.e. located on a spiral). The resulting series of transverse axes permit the sliding and rolling motion in the flexing knee while ensuring the collateral ligaments are sufficiently lax to permit the rotation associated with the curvature of the medial condyle about a vertical axis.

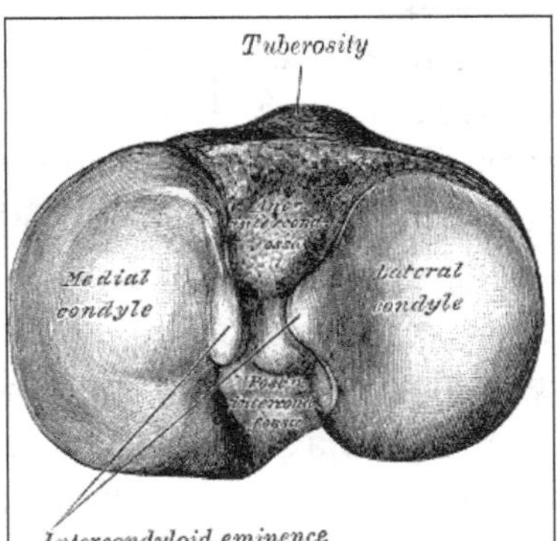

Articular surfaces of tibia.

The pair of tibial condyles are separated by the intercondylar eminence composed of a lateral and a medial tubercle.

The patella also serves an articular body, and its posterior surface is referred to as the trochlea of the knee. It is inserted into the thin anterior wall of the joint capsule. On its posterior surface is a lateral and a medial articular surface, both of which communicate with the patellar surface which unites the two femoral condyles on the anterior side of the bone's distal end.

Articular Capsule

The articular capsule has a synovial and a fibrous membrane separated by fatty deposits.

Anteriorly, the synovial membrane is attached on the margin of the cartilage both on the femur and the tibia, but on the femur, the suprapatellar bursa or recess extends the joint space proximally. The suprapatellar bursa is prevented from being pinched during extension by the articularis genus muscle. Behind, the synovial membrane is attached to the margins of the two femoral condyles which produces two extensions similar to the anterior recess. Between these two extensions, the synovial membrane passes in front of the two cruciate ligaments at the center of the joint, thus forming a pocket direct inward.

Bursae

Numerous bursae surround the knee joint. The largest communicative bursa is the suprapatellar bursa described above. Four considerably smaller bursae are located on the back of the knee. Two non-communicative bursae are located in front of the patella and below the patellar tendon, and others are sometimes present.

Cartilage

Cartilage is a thin, elastic tissue that protects the bone and makes certain that the joint surfaces can slide easily over each other. Cartilage ensures supple knee movement. There are two types of joint cartilage in the knees: fibrous cartilage (the meniscus) and hyaline cartilage. Fibrous cartilage has tensile strength and can resist pressure. Hyaline cartilage covers the surface along which the joints move. Collagen fibres within the articular cartilage have been described by Benninghoff as arising from the subchondral bone in a radial manner, building so called Gothic arches. On the surface of the cartilage these fibres appear in a tangential orientation and increase the abrasion resistance. There are no blood vessels inside of the hyaline cartilage, the alimentation is performed per diffusion. Synovial fluid and the subchondral bone marrow serve both as nutrition sources for the hyaline cartilage. Lack of at least one source induces a degeneration. Cartilage will wear over the years. Cartilage has a very limited capacity for self-restoration. The newly formed tissue will generally consist of a large part of fibrous cartilage of lesser quality than the original hyaline cartilage. As a result, new cracks and tears will form in the cartilage over time.

Menisci

The articular disks of the knee-joint are called menisci because they only partly divide the joint space. These two disks, the medial meniscus and the lateral meniscus, consist of connective tissue with extensive collagen fibers containing cartilage-like cells. Strong fibers run along the menisci from one attachment to the other, while weaker radial fibers are interlaced with the former. The menisci are flattened at the center of the knee joint, fused with the synovial membrane laterally, and can move over the tibial surface.

The menisci serve to protect the ends of the bones from rubbing on each other and to effectively deepen the tibial sockets into which the femur attaches. They also play a role in shock absorption, and may be cracked, or torn, when the knee is forcefully rotated and bent.

Ligaments

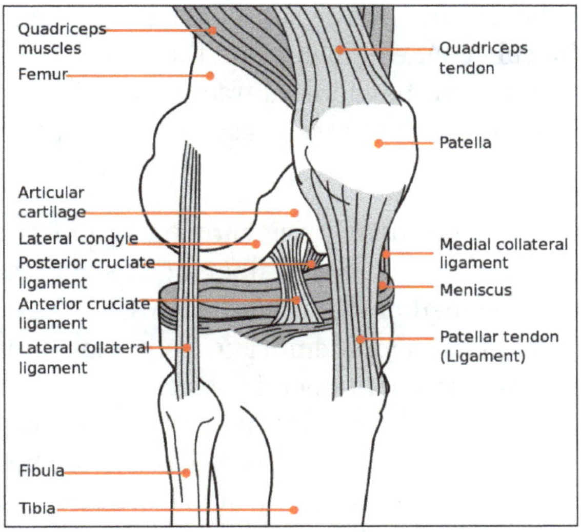

Anterolateral aspect of right knee.

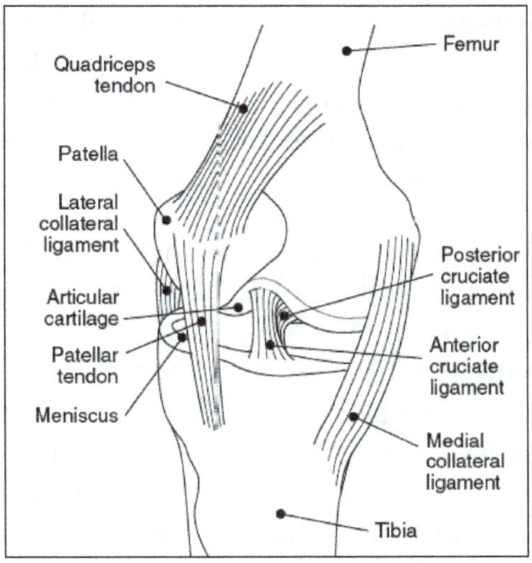

Anteromedial aspect of right knee.

The ligaments surrounding the knee joint offer stability by limiting movements and, together with the menisci and several bursae, protect the articular capsule.

Intracapsular

The knee is stabilized by a pair of cruciate ligaments. The anterior cruciate ligament (ACL) stretches from the lateral condyle of femur to the anterior intercondylar area. The ACL is critically important because it prevents the tibia from being pushed too far anterior relative to the femur. It is often torn during twisting or bending of the knee. The posterior cruciate ligament (PCL) stretches from medial condyle of femur to the posterior intercondylar area. Injury to this ligament is uncommon but can occur as a direct result of forced trauma to the ligament. This ligament prevents posterior displacement of the tibia relative to the femur.

The transverse ligament stretches from the lateral meniscus to the medial meniscus. It passes in front of the menisci. It is divided into several strips in 10% of cases. The two menisci are attached to each other anteriorly by the ligament. The posterior and anterior meniscofemoral ligaments stretch from the posterior horn of the lateral meniscus to the medial femoral condyle. They pass posteriorly behind the posterior cruciate ligament. The posterior meniscofemoral ligament is more commonly present (30%); both ligaments are present less often. The meniscotibial ligaments (or "coronary") stretches from inferior edges of the mensici to the periphery of the tibial plateaus.

Extracapsular

The patellar ligament connects the patella to the tuberosity of the tibia. It is also occasionally called the patellar tendon because there is no definite separation between the quadriceps tendon (which surrounds the patella) and the area connecting the patella to the tibia. This very strong ligament helps give the patella its mechanical leverage and also functions as a cap for the condyles of the femur. Laterally and medially to the patellar ligament the lateral and medial retinacula connect fibers from the vasti lateralis and medialis muscles to the tibia. Some fibers from the iliotibial tract radiate into the lateral retinaculum and the medial retinaculum receives some transverse fibers arising on the medial femoral epicondyle.

The medial collateral ligament stretches from the medial epicondyle of the femur to the medial tibial condyle. It is composed of three groups of fibers, one stretching between the two bones, and two fused with the medial meniscus. The MCL is partly covered by the pes anserinus and the tendon of the semimembranosus passes under it. It protects the medial side of the knee from being bent open by a stress applied to the lateral side of the knee (a valgus force). The fibular collateral ligament stretches from the lateral epicondyle of the femur to the head of fibula. It is separate from both the joint capsule and the lateral meniscus. It protects the lateral side from an inside bending force (a varus force). The anterolateral ligament (ALL) is situated in front of the LCL.

Lastly, there are two ligaments on the dorsal side of the knee. The oblique popliteal ligament is a radiation of the tendon of the semimembranosus on the medial side, from where it is direct laterally and proximally. The arcuate popliteal ligament originates on the apex of the head of the fibula to stretch proximally, crosses the tendon of the popliteus muscle, and passes into the capsule.

Calf Leg

The calf is the back portion of the lower leg in human anatomy. The muscles within the calf correspond to the posterior compartment of the leg. The two largest muscles within this compartment are known together as the calf muscle and attach to the heel via the Achilles tendon. Several other, smaller muscles attach to the knee, the ankle, and via long tendons to the toes.

Structure

The calf is composed of the muscles of the posterior compartment of the leg: The gastrocnemius and soleus (composing the triceps surae muscle) and the tibialis posterior. The sural nerve provides innervation.

Foot

The foot can be divided into three sections: the forefoot, midfoot, and hindfoot. There are bones, joints, muscles, tendons and ligaments in each section:

- Bones: Bones in the foot can be divided into three categories based on where they are located.
- Forefoot: This is the very front part of the foot, including the toes, or phalanges. There are 14 toe bones (two per big toe and three per each of the other four), plus five metatarsals. The first metatarsal bone is the shortest and thickest and plays an important role during propulsion (forward movement). It also provides attachment to several tendons. The second, third, and fourth metatarsal bones are the most stable of the metatarsals. They are well-protected and have only minor tendon attachments.
- In addition to the phalanges and metatarsals, the forefoot contains two small, oval-shaped sesamoid bones just beneath the head of the first metatarsal, on the plantar surface, or underside, of the foot, that is held in place by tendons and ligaments. The forefoot meets the midfoot at the five tarsometatarsal (TMT) joints.
- Midfoot: This section of the foot is made up of five irregularly-shaped bones called the tarsals. The clinical names for these bones are the navicular, cuboid, and medial, intermediate, and lateral cuneiforms. Together they form the arch of the foot. The arch of the foot plays a key role in weight-bearing and stability.
- Hindfoot: There are only two large bones in this section of the foot: the talus and the calcaneus. The largest of these, the calcaneus, forms the heel of the foot. The talus rests on top of the calcaneus and forms the pivoting joint of the ankle.
- Joints: A joint is formed at the junction between two or more bones. Each big toe has two joints, the metatarsophalangeal joint, and the interphalangeal joint. The other four toes on each foot have three joints each: the metatarsophalangeal joint (MCP) at the base of the toe, the proximal interphalangeal joint (PIP) in the middle of the toe, and the distal phalangeal joint (DP)—the joint closest to the tip of the toe.
- Muscles: The muscles that control the movements of the foot originate in the lower leg and are attached the bones in the foot with tendons. The main muscles that facilitate movement in the foot are:
    - Tibialis posterior (supports the foot's arch).
    - Tibialis anterior (allows the foot to move upward).
    - Tibialis peroneal (controls movement on the outside of the ankle).

- Extensors (raise the toes to make possible to take a step).
- Flexors (stabilize the toes).

- Tendons and Ligaments: The most notable tendon of the foot is the Achilles tendon, which runs from the calf muscle to the heel. It is the strongest and largest tendinous structure in the body. The Achilles tendon makes it possible to run, jump, climb stairs, and stand on your toes. Other important tendons in the foot include the tibialis posterior (posterior tibial tendon), which attaches the calf muscle to the bones on the inside of the foot and supports the arch of the foot, and the tibialis anterior (anterior tibial tendon) which runs from the outer tibia to the first metatarsal and surfaces of the median cuneiform tarsal, which allows for dorsiflexion—bringing the toes toward the shin.

The primary ligaments of the foot are the:

- Plantar fascia: The longest ligament of the foot, the plantar fascia runs along the sole of the foot from the heel to the toes to form the arch of the foot, provide strength for walking, and to assist with balance.
- Plantar calcaneonavicular ligament: This is a ligament of the sole of the foot that connects the calcaneus and navicular and supports the head of the talus.
- Calcaneocuboid ligament: This is the ligament that connects the calcaneus and the tarsal bones and helps the plantar fascia support the arch of the foot.

Ankle

The ankle is actually made up of two joints: the subtalar joint and the true ankle joint. The true ankle joint is composed of three bones, seen above from a front, or anterior, view: the tibia which forms the inside, or medial, portion of the ankle; the fibula which forms the lateral, or outside portion of the ankle; and the talus underneath. The true ankle joint is responsible for the up-and-down motion of the foot.

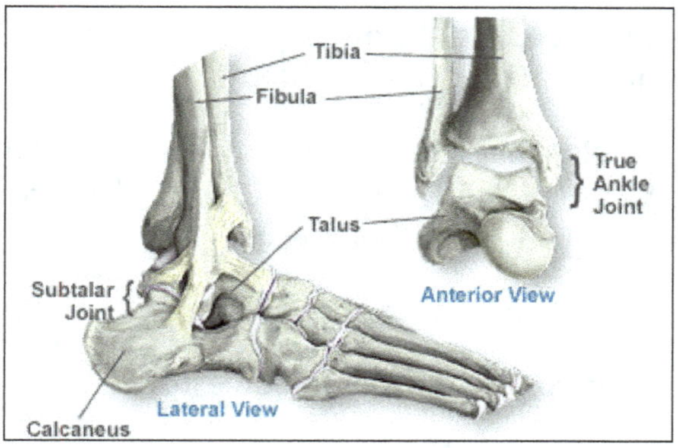

Beneath the true ankle joint is the second part of the ankle, the subtalar joint, which consists of the talus on top and calcaneus on the bottom. The subtalar joint allows side-to-side motion of the foot.

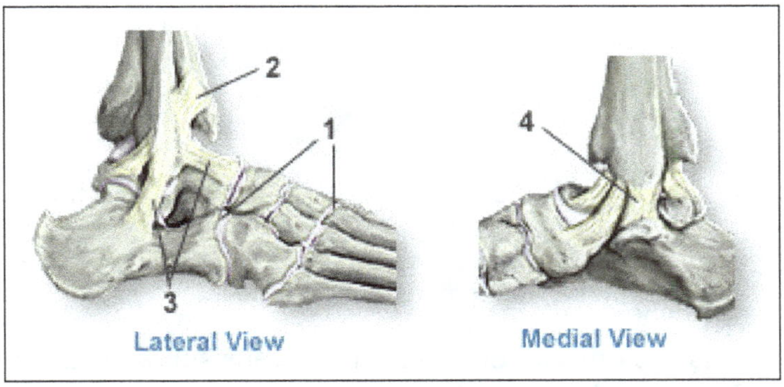

The ends of the bones in these joints are covered by articular cartilage. The major ligaments of the ankle are:

- The anterior tibiofibular ligament, which connects the tibia to the fibula.
- The lateral collateral ligaments, which attach the fibula to the calcaneus and gives the ankle lateral stability.
- On the medial side of the ankle, the deltoid ligaments, which connect the tibia to the talus and calcaneus and provide medial stability.

These components of your ankle, along with the muscles and tendons of your lower leg, work together to handle the stress your ankle endures as you walk, run, and jump.

Heel

The heel is the prominence at the posterior end of the foot. It is based on the projection of one bone, the calcaneus or heel bone, behind the articulation of the bones of the lower leg.

Anatomical Structure

To distribute the compressive forces exerted on the heel during gait, and especially the stance phase when the heel contacts the ground, the sole of the foot is covered by a layer of subcutaneous connective tissue up to 2 cm thick (under the heel). This tissue has a system of pressure chambers that both acts as a shock absorber and stabilises the sole. Each of these chambers contains fibrofatty tissue covered by a layer of tough connective tissue made of collagen fibers. These septa ("walls") are firmly attached both to the plantar aponeurosis above and the sole's skin below. The sole of the foot is one of the most highly vascularized regions of the body surface, and the dense system of blood vessels further stabilize the septa.

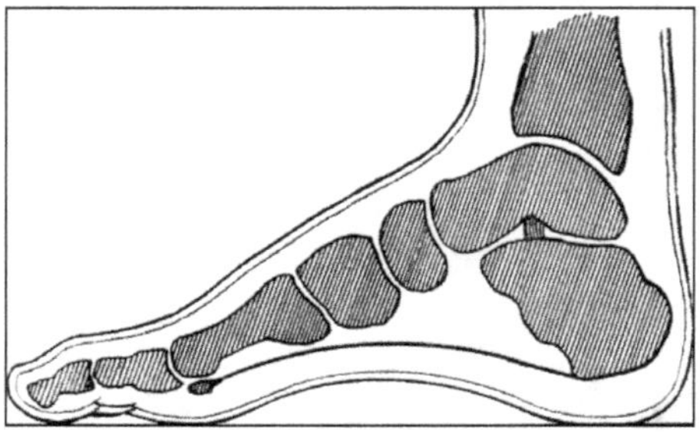

Sagittal section through the foot.

The Achilles tendon is the muscle tendon of the triceps surae, a “three-headed” group of muscles—the soleus and the two heads of the gastrocnemius. The main function of the triceps surae is plantar flexion, i.e. to stretch the foot downward. It is accompanied by a “fourth head”, the slight plantaris muscle, the long slender tendon of which is also attached to the heel bone but not visible.

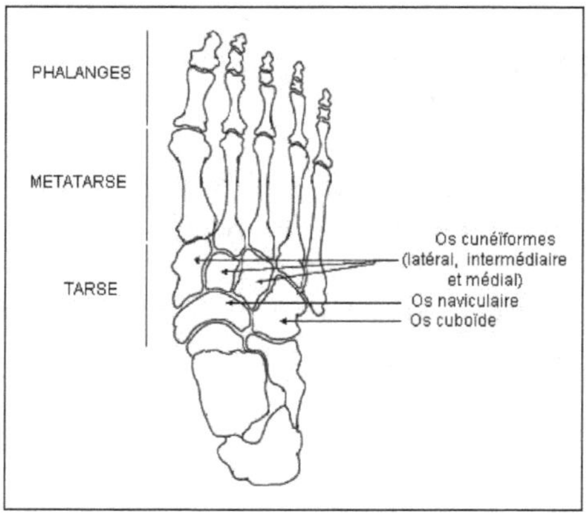

From above.

Toe

Toes are the digits (fingers) of the foot of a tetrapod. Animal species such as cats that walk on their toes are described as being digitigrade. Humans, and other animals that walk on the soles of their feet, are described as being plantigrade; unguligrade animals are those that walk on hooves at the tips of their toes.

Structure

There are five toes present on each human foot. Each toe consists of three phalanx bones, the proximal, middle and distal, with the exception of the big toe. For a minority of people, the little toe also is missing a middle bone. The hallux only contains two phalanx bones, the proximal and distal. The joints between each phalanx are the interphalangeal joints. The proximal phalanx bone of each toe articulates with the metatarsal bone of the foot at the metatarsophalangeal joint. Each toe is surrounded by skin, and present on all five toes is a toenail.

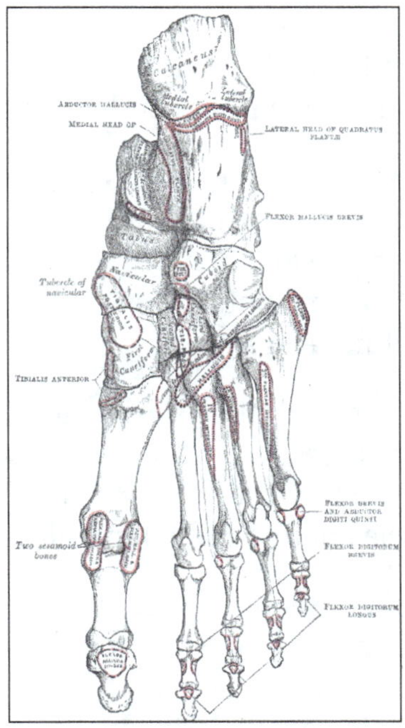

Bones of the right foot - Plantar surface.

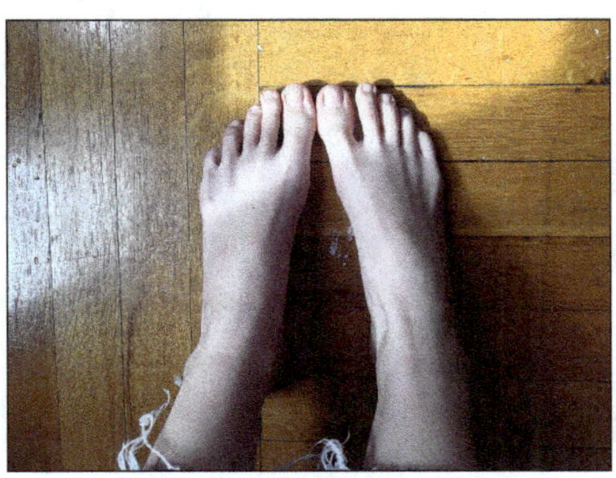
Female toes.

The toes are, from medial to lateral:

- The first toe, also known as the hallux ("big toe", "great toe", "thumb toe"), the innermost toe.
- The second toe, ("long toe", "pointer toe", "index toe").
- The third toe, ("middle toe", "long toe").

- The fourth toe, ("fore toe").
- The fifth toe, ("baby toe", "little toe", "small toe", "tiny toe", "wee toe", "pinky toe"), the outermost toe.

Muscles

Toe movement is generally flexion and extension via muscular tendons that attach to the toes on the anterior and superior surfaces of the phalanx bones.

With the exception of the hallux, toe movement is generally governed by action of the flexor digitorum brevis and extensor digitorum brevis muscles. These attach to the sides of the bones, making it impossible to move individual toes independently. Muscles between the toes on their top and bottom also help to abduct and adduct the toes. The hallux and little toe have unique muscles:

- The hallux is primarily flexed by the flexor hallucis longus muscle, located in the deep posterior of the lower leg, via the flexor hallucis longus tendon. Additional flexion control is provided by the flexor hallucis brevis. It is extended by the abductor hallucis muscle and the adductor hallucis muscle.
- The little toe has a separate set of control muscles and tendon attachments, the flexor and abductor digiti minimi. Numerous other foot muscles contribute to fine motor control of the foot. The connective tendons between the minor toes account for the inability to actuate individual toes.

Blood Supply

The toes receive blood from the digital branches of the plantar metatarsal arteries and drain blood into the dorsal venous arch of the foot.

Nerve Supply

Sensation to the skin of the toes is provided by five nerves. The superficial fibular nerve supplies sensation to the top of the toes, except between the hallux and second toe, which is supplied by the deep fibular nerve, and the outer surface of the fifth toe, supplied by the sural nerve. Sensation to the bottom of the toes is supplied by the medial plantar nerve, which supplies sensation to the great toe and inner three-and-a-half toes, and the lateral plantar nerve, which supplies sensation to the little toe and half of the sensation of the fourth toe.

In humans, the hallux is usually longer than the second toe, but in some individuals, it may not be the longest toe. There is an inherited trait in humans, where the dominant gene causes a longer second toe while the homozygous recessive genotype presents with the more common trait: a longer hallux. People with the rare genetic disease fibrodysplasia ossificans progressiva characteristically have a short hallux which appears to turn inward, or medially, in relation to the foot.

Variation

Humans usually have five toes on each foot. When more than five toes are present, this is known as polydactyly. Other variants may include syndactyly or arachnodactyly. Forefoot shape, including toe shape, exhibits significant variation among people; these differences can be measured and have been statistically correlated with ethnicity. Such deviations may affect comfort and fit for various shoe types.

Sole

The sole is the bottom of the foot. In humans the sole of the foot is anatomically referred to as the plantar aspect.

Structure

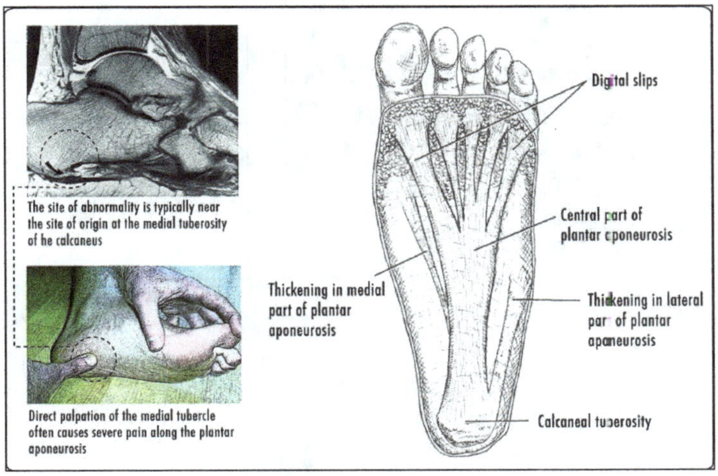

Deep anatomy of the sole.

The glabrous skin on the sole of the foot lacks the hair and pigmentation found. The glabrous skin on the sole of the foot lacks the hair and pigmentation found elsewhere on the body, and it has a high concentration of sweat pores. The sole contains the thickest layers of skin on the body due to the weight that is continually placed on it. It is crossed by a set of creases that form during the early stages of embryonic development. Like those of the palm, the sweat pores of the sole lack sebaceous glands.

The sole is a sensory organ by which we can perceive the ground while standing and walking. The subcutaneous tissue in the sole has adapted to deal with the high local compressive forces on the heel and the ball (between the toes and the arch) by developing a system of "pressure chambers." Each chamber is composed of internal fibrofatty tissue covered by external collagen connective tissue. The septa (internal walls) of these chambers are permeated by numerous blood vessels, making the sole one of the most vascularized, or blood-enriched, regions in the human body.

The sole and the longitudinal arch of the foot are supported by a thick connective tissue, the plantar fascia. The central component of this tissue extends to the supporting bones and gives two divisions–the medial component and lateral component; thus they define the boundaries of the three muscle compartments of the sole.

The bones underlying the sole form the arch of the foot. The arches might collapse later in life, resulting in flat feet.

Muscles

Intrinsic

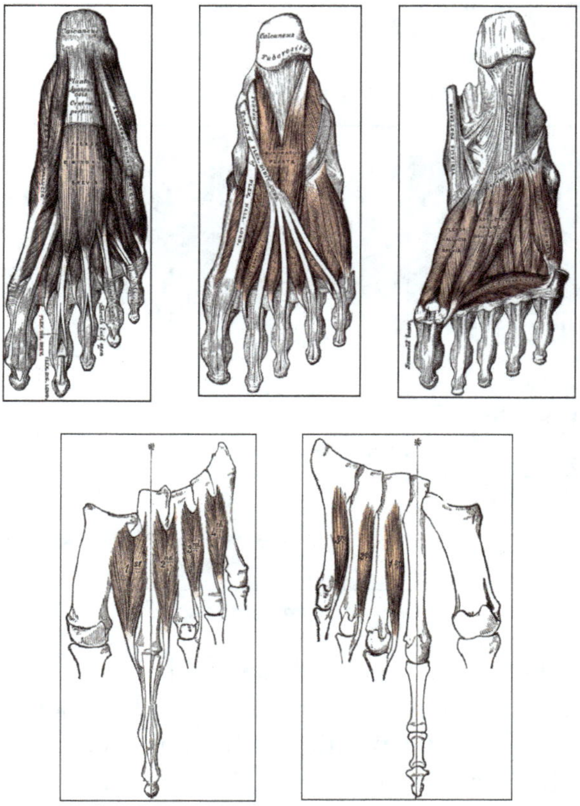

First, second, and third muscle layers, and the dorsal and plantar interossei.

The intrinsic muscles in the sole are grouped in four layers:

- In the first layer: The flexor digitorum brevis is the large central muscle located immediately above the plantar aponeurosis. It flexes the second to fifth toes and is flanked by abductor hallucis and abductor digiti minimi.

- In the second layer: The quadratus plantae, located below flexor digitorum brevis, inserts into the tendon of flexor digitorum longus on which the lumbricals originate.
- In the third layer: The oblique head of adductor hallucis joins the muscle's transversal head on the lateral side of the big toe. Medially to adductor hallucis are the two heads of flexor hallucis brevis, deep to the tendon of flexor hallucis longus. The considerably smaller flexor digiti minimi brevis on the lateral side can be mistaken for one of the interossei.
- In the fourth layer: The dorsal and plantar interossei are located between and below the metatarsal bones and act as antagonists.

The central compartment is shared by the lumbricals, quadratus plantae, flexor digitorum brevis, and adductor hallucis; the medial compartment by abductor hallucis, flexor hallucis brevis, abductor digiti minimi, flexor digiti minimi brevis, and opponens digiti minimi (often considered part of the former muscle); whilst the lateral compartment is occupied by extensor digitorum brevis and extensor hallucis brevis.

Extrinsic

The tendons of several extrinsic foot muscle reach the sole:

- The tendons of the deep foot flexors in the posterior compartment of the leg, tibialis posterior, flexor digitorum longus, and flexor hallucis longus, passes behind the medial malleolus into the sole.
- The tendon of fibularis longus similarly passes behind the lateral malleolus into the sole.

Nerve Supply

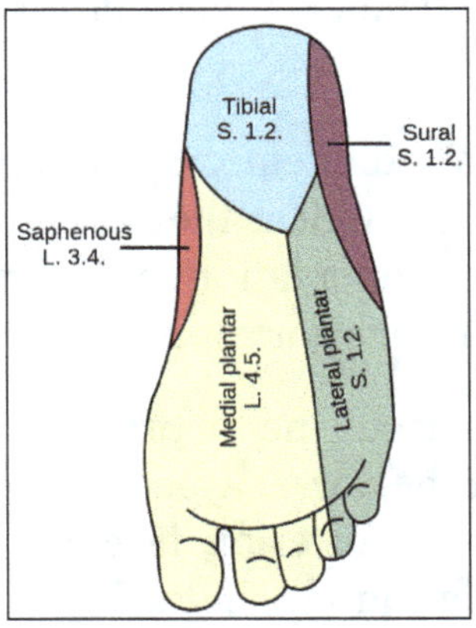

Cutaneous innervation of the sole of the foot.

- The soles of the feet are extremely sensitive to touch due to a high concentration of nerve endings, with as many as 200,000 per sole. This makes them sensitive to surfaces that are walked on, ticklish and some people find them to be erogenous zones.
- Medically, the soles are the site of the plantar reflex, the testing of which can be painful due to the sole's sensitivity.
- The deep fibular nerve from the common fibular nerve provides the sensory innervation of the skin between the first and second toes and the motor innervation of the muscles of the anterior compartment of the leg and dorsal foot. Damage to the deep fibular nerve can result in foot drop.
- The plantar digital nerves from the medial plantar nerve provide sensory innervation to the skin of the plantar aspect of the toes, except the medial part of the big toe and the lateral part of the little toe and the motor innervation of the first lumbrical.
- The proper plantar nerve from the common plantar digital nerve provide sensory innervation to the plantar surface of the toes as well as the dorsal aspect of the distal interphalangeal phalanges. It also provides motor innervation to flexor hallucis brevis.
- The superficial and deep branches of the lateral plantar nerve from the tibial nerve provide sensory innervation to the skin of the lateral side of the sole, to the fifth and half the fourth toes, and the nail bed of these toes. They also provide motor innervation to quadratus plantae, abductor digiti minimi, flexor digiti minimi brevis, lateral three lumbricals, adductor hallucis, and the dorsal and plantar interossei.
- The medial plantar nerve from the tibial nerve provides sensory innervation to the skin of the medial side of the sole, the skin of the medial three and a half toes, and the nail beds of these toes. It also provides motor innervation to abductor hallucis, flexor hallucis brevis, flexor digitorum brevis, and the first lumbrical.
- The saphenous nerve from the femoral nerve provides sensory innervation to the medial side of the foot as well as the medial side of the leg. Likewise, the sural nerve provides sensory innervation to the skin on the lateral side of the foot as well as the skin on the posterior aspect of the lower leg.
- The tibial nerve from the sciatic nerve provides sensory innervation to the skin of the sole and toes, and the dorsal aspect of the toes. It provides motor innervation to plantaris, tibialis posterior, flexor hallucis longus, flexor digitorum longus as well as posterior muscles in the leg.

Muscles of the Lower Limb

Muscles that cause Movement at the Hip Joint

The four main groups of hip muscles are gluteal, adductor, iliopsoas, and lateral rotator, defined by the type of movement they mediate.

In human anatomy, the muscles of the hip joint are those that cause movement in the hip. Hip joint muscles are divided into four groups according to their orientation and function. Movement at the hip is similar to that of the shoulder joint, but due to increased weight-bearing requirements the range of potential movements is reduced.

Gluteal Group

Muscles in the gluteal group are superficially located and act mainly to abduct and extend the thigh at the hip.

- Gluteus Maximus: The gluteus maximus is the largest of the gluteal muscles and gives structure to the buttocks.
    - Attachments: Originates from the posterior of the pelvis and coccyx (tailbone) and attaches to the femur.
    - Actions: Extends of the thigh and assists with rotation. Is only used when the generation of force is required (e.g. when climbing).
- Gluteus Medius: The fan-shaped gluteus medius muscle lies between the gluteus maximus and minimus and performs a similar function to the gluteus minimus.
    - Attachments: Originates from the posterior of the pelvis and attaches to the femur.
    - Actions: Abducts and medially rotates the thigh and fixes the pelvis during walking.
- Gluteus Minimus: The gluteus minimus is the deepest and smallest of the superficial gluteal muscles and performs a similar function to the gluteus medius.
    - Attachments: Originates from the pelvis and attaches to the femur.
    - Actions: Abducts and medially rotates the thigh and fixes the pelvis during walking.

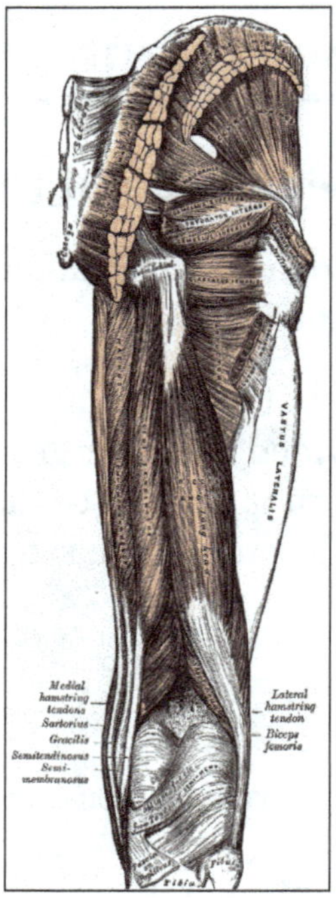

The gluteus maximus can be seen at the top, cut away to expose the underlying muscles.

Lateral Rotator Group

The muscles of the lateral rotator group are deeply located and as the name suggests, act to laterally rotate the thigh at the hip. All of the lateral rotator group muscles originate from the pelvis and attach to the femur.

- Piriformis: The piriformis is the most superior of the lateral rotator group muscles.
    - Actions: Lateral rotation and abduction of the thigh at the hip.
- Obturator Internus: The obturator internus lines the internal wall of the pelvis.
    - Actions: Lateral rotation and abduction of the thigh at the hip.
- Gemelli: The gemelli are two (superior and inferior) narrow and triangular muscles, separated by the obturator internus tendon.
    - Actions: Lateral rotation and abduction of the thigh at the hip.
- Quadratus Femoris: The quadratus femoris is a flat, square-shaped muscle (actually composed of four distinct muscles). It is the most inferior of the lateral rotator group muscles, located below the gemelli and obturator internus.
    - Actions: Lateral rotation of the thigh at the hip, plays a major role in extension of the lower leg at the knee as well.

Adductor Group

The five muscles of the adductor group are responsible for the adduction of the thigh, although several have additional functions.

- Adductor Longus: The adductor longus is a large, flat muscle covering the adductor magnus and adductor brevis.
    - Attachments: Originates from the pubis and broadly attaches to the femur.
    - Actions: Adduction and medial rotation of the thigh.

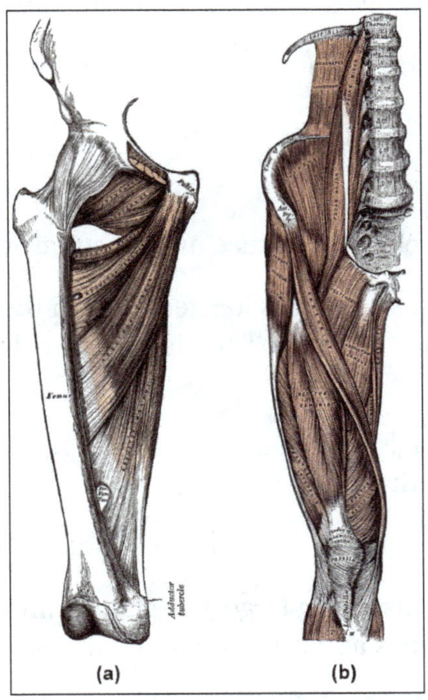

(a) Adductor Group Muscles and (b) Key muscles associated with movement at the hip: The deep-lying adductor group muscles originate from the pubis and attach to the length of the femur. The iliacus and psoas major comprise the iliopsoas group and can be seen at the pelvis and lower spine.

- Adductor Magnus: The adductor magnus is the largest and most posterior of the adductor group muscles.
    - Attachments: Originates from the pubis and attaches to the femur.
    - Actions: Adducts, flexes, and extends the thigh.
- Adductor Brevis: The adductor brevis is a short muscle lying underneath the adductor longus.
    - Attachments: Originates from the pubis and attaches to the femur.
    - Actions: Adduction of the thigh.

- Obturator Externus: This is one of the smaller muscles of the medial thigh, and it is located most superiorly.
    - Attachments: Originates from the pubis and attaches to the femur.
    - Actions: Laterally rotates the thigh.
- Gracilis: The gracilis is the most superficial and medial of the adductor group muscles. Crossing both the hip and knee joints, it can induce movement at both the hip and knee.
    - Attachments: Originates from the pubis and attaches to the tibia.
    - Actions: Adduction of the thigh at the hip, and flexing of the thigh at the knee.

Other Muscles

There are several other muscles that induce movement around the hip joint.

- Psoas Major: The psoas major is located deep in the back near the midline immediately adjacent to the spine. The iliacus and psoas major comprise the iliopsoas group.
    - Attachments: Originates from the base of the spine, combining with the iliacus to attach to the femur.
    - Actions: Flexing of the thigh at the hip joint.
- Iliacus: The iliacus muscle is a large, fan-shaped muscle which lines the interior of the pelvis. The iliacus and psoas major comprise the iliopsoas group.
    - Attachments: Originates from the pelvis and the base of the spine, combining with the psoas major to attach to the femur.
    - Actions: Flexing of the thigh at the hip joint.
- Sartorius: The sartorius is a long thin muscle in the thigh, the longest muscle in the body.
    - Attachments: Originates from the pelvis and attaches to the tibia.
    - Actions: Flexing, abducting and rotation of the thigh at the hip joint.
- Pectineus: The pectineus muscle is a large flat muscle found in the thigh.
    - Attachments: Originates from the pelvis and attaches to the femur.
    - Actions: Adduction and flexing at the thigh at the hip joint.

- Biceps Femoris: A similar muscle to the biceps brachii in the upper arm, also double-headed. Two synergistic muscles are associated with the biceps femoris, the semitendinosus, and the semimembranosus:
    - Attachments: Originates from the pelvis and femur and attaches to the fibula.
    - Actions: Extends and laterally rotates at the hip. The main action is flexing of the lower leg at the knee.

Muscles that cause Movement at the Knee Joint

Three sets of muscles (popliteus, quadriceps and hamstrings) allow for movement, balance, and stability at the knee joint.

The knee joint allows for movement of the lower leg relative to the thigh across the knee joint. The knee joint is in fact comprised of two joints: the tibiofemoral joint between the femur and tibia, which is the weight-bearing knee joint, and the patellofemoral joint, which joins the patella (kneecap) with the femur.

The tibiofemoral joint is relatively weak and easily damaged, so it relies on muscles and ligaments to ensure stability. When the knee is fully extended the femur rotates slightly on the tibia to lock the joint into place, allowing for efficient load bearing.

The patella is the attachment point for the quadriceps femoris muscle and is the attached by a ligament to the tibia. This increases the leverage afforded to the quadriceps femoris muscle, thus increasing its efficiency when extending the lower leg. The patella additionally protects the knee joint from damage.

The patellofemoral has two key functions: increasing leverage of the quadriceps tendon to improve muscle stability and protecting the knee joint from damage.

Muscles that generate movement across the knee are mainly located in the thigh and can be split into anterior and posterior compartments. The popliteus muscle, located in the lower leg, is responsible for "unlocking" the knee joint after extension.

Anterior Muscles of the Thigh

There are four muscles in anterior region of the thigh. The pectineus and iliopsoas muscles are responsible for movement at the hip and are discussed elsewhere:

- Sartorius: The sartorius, a thin muscle in the thigh, this is the body's longest muscle.
    - Attachments: Originates from the pelvis and attaches to the tibia.
    - Actions: Flexing of the lower leg at the knee joint.

- Quadriceps Femoris: The quadriceps femoris is actually composed of four muscles that comprise the front of the thigh: three deep-lying vastus muscles (lateralis, intermedius, and medialis) and the rectus femoris which covers them. All four muscles are the key extensors of the lower leg at the knee joint and also stabilize and protect the patella.
    - Attachments: The vastus lateralis, intermedius, and medialis originate from the femur and attach to the patella. The rectus femoris originates from the pelvis and attaches to the patella.
    - Actions: Extends the lower leg at the knee joint and stabilizes the patella. The rectus femoris additionally facilitates rotation at the hip.

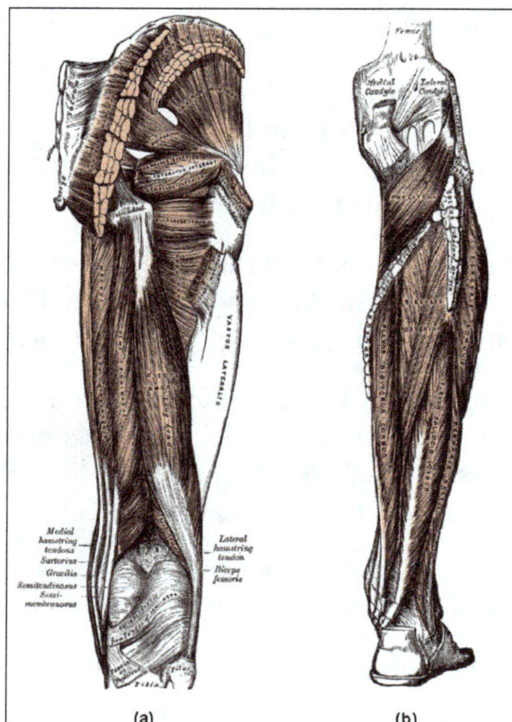

(a) Posterior muscles of the thigh and (b) posterior region of the lower leg: The biceps femoris and synergistic semitendinosus and the semimembranosus muscles are responsible for flexing of the lower leg at the knee. Posterior view of muscles of the lower leg, the popliteus can be seen at the top located behind the knee.

Posterior Muscles of the Thigh

There are three muscles in the posterior compartment of the thigh: the biceps femoris and two synergistic muscles (the semitendinosus and semimembranosus). These muscles are sometimes termed the hamstring group. The posterior region of the thigh displays similarity with the anterior region of the upper arm in both structure and function.

- Biceps Femoris: A similar muscle to the biceps brachii in the upper arm and also double-headed. Two synergistic muscles are associated with the biceps femoris, the semitendinosus and the semimembranosus.
    - Attachments: Originates from the pelvis and femur and attaches to the fibula.
    - Actions: Extends and laterally rotates at the hip, main action is flexing of the lower leg at the knee.

Other Muscles

- Popliteus: The popliteus is located behind the knee joint and acts to “unlock” the knee by rotating the femur on the tibia allowing for the lower leg to be flexed.
    - Attachments: Originates from the posterior of the tibia and attaches to the femur.
    - Actions: Laterally rotates the femur on the tibia “unlocking” the knee joint so that flexion can occur.

Muscles that cause Movement at the Ankle

Muscles of the leg insert into ankle and foot bones to facilitate ankle movement. Movement at the ankle is controlled by two joints. The ankle or talocrural joint is formed from the tibia and fibula of the lower leg and talus of the foot. Functionally, it acts as a hinge, allowing dorsiflexion (pulling the foot upwards towards the lower leg) and plantarflexion (pulling the foot downwards away from the lower leg). Eversion (tilting of the sole of the foot away from the midline) and inversion (tilting of the sole of the foot inwards towards the midline) is controlled by the subtalar joint formed between the talus and calcaneus bones of the foot.

The ankle joint is held in place by numerous strong ligaments that can be easily damaged when excessive force is placed on the ankle, particularly during strenuous inversion and eversion. Movement at the ankle is key for maintenance of posture and balance, but is most important in locomotion. Variation in muscle activation can control the movement of the ankle joint, allowing the foot to generate graduated force.

Muscles that generate movement at the ankle are generally found in the lower leg and can be split into three categories.

Anterior Compartment

Three muscles in the anterior compartment of the leg act to dorsiflex and invert the foot at the ankle joint.

- Tibialis Anterior: The tibialis anterior muscle is located alongside the lateral surface of the tibia and is the strongest dorsiflexor of the foot.

   - Attachments: Originates from the lateral surface of the tibia and attaches to the base of the big toe.
   - Actions: Dorsiflexion and inversion of the foot.
- Extensor Digitorum Longus: The extensor digitorum longus is a deep-lying extrinsic muscle that runs the length of the tibia.
   - Attachments: Originates from the tibia and transitions into a tendon, passes into the foot, splits into four, and attaches to the toes.
   - Actions: Extension of the toes and dorsiflexion of the foot.
- Extensor Hallucis Longus: The extensor hallucis longus is a deep lying extrinsic muscle beneath the extensor digitorum longus.
   - Attachments: Originates from the fibula and attaches to the big toe.
   - Actions: Extension of the big toe, and dorsiflexion of the foot.

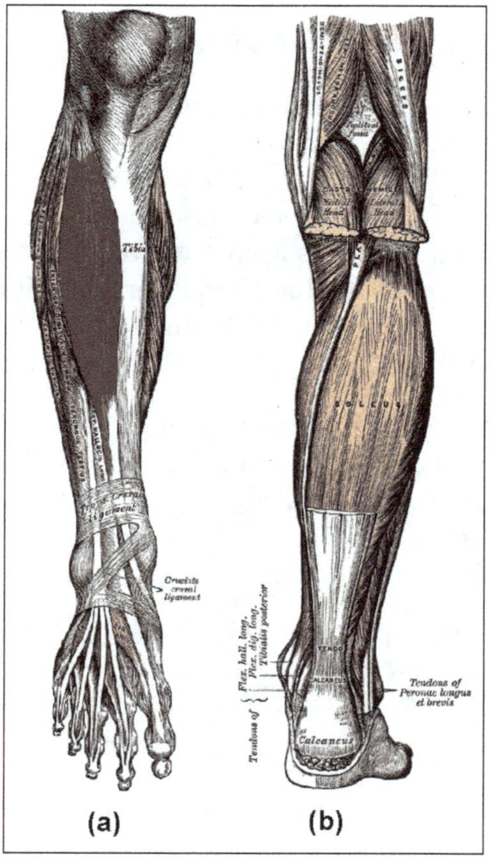

(a) Anterior Compartment of the Leg and (b) Posterior Compartment of the leg: Anterior view of leg showing the muscles and tendons involved in ankle movement. Posterior view of leg showing muscles and tendons involved in ankle movement.

Posterior Compartment

Several muscles are located in the posterior compartment of the leg, typically grouped into superficial and basal groups. The majority of these muscles work to plantarflex the foot at the ankle.

Superficial Muscles

The superficial muscles give rise to the characteristic shape of the lower leg.

- Gastrocnemius: The gastrocnemius, a two-headed muscle, is the most superficial of the muscles in the posterior compartment.
    - Attachments: Both heads originate from the femur. The fibers converge to form the calcaneal tendon which attaches to the heel.
    - Actions: Plantarflexes the foot, can also flex the lower leg at the knee but is not key in this movement.
- Plantaris: The plantaris is a small muscle lying between the gastrocnemius and soleus. It is absent in 10% of people.
    - Attachments: Originates from the femur and attaches to the heel via the calcaneal tendon.
    - Actions: Plantarflexes the foot, can also flex the lower leg at the knee but is not key in this movement.
- Soleus: The soleus is a large flat muscle which is the deepest lying of the superficial muscles.
    - Attachments: Originates from the tibia and fibula and attaches to the heel via the calcaneal tendon.
    - Actions: Plantarflexes the foot.

Deep Muscles

- Tibialis Posterior: The tibialis posterior is the deepest lying of the muscles in the posterior compartment.
    - Attachments: Originates from the tibia and fibula and attaches to the plantar surfaces of the toes.
    - Actions: Inverts and plantarflexes the foot, maintains the arch of the foot.

Lateral Compartment

Two muscles found in the lateral compartment function to control eversion of the foot.

Physiologically, there is a preference for the foot to invert, so these muscles also prevent excessive inversion.

- Fibularis Longus: The fibularis longus is the longer and more superficial of the two muscles.
  - Attachments: Originates from the fibula and tibia. The fibers converge into a tendon which passes under the foot and attaches to the medial side of the foot.
  - Actions: Eversion and plantarflexion of the foot.
- Fibularis Brevis: The fibularis brevis muscle is the deeper and shorter of the two muscles.
  - Attachments: Originates from the lateral surface of the fibula and attaches to the little toe.
  - Actions: Eversion of the foot.

Muscles that cause Movement at the Foot

Movement of the foot and toes requires the action of many muscles.

The foot plays a fundamental role in maintaining upright walking posture and balance, so it must provide a strong base while allowing for small, accurate movements. The foot can be split into dorsal and plantar compartments, the latter of which contains significantly more muscles.

Dorsal Compartment

Although many extrinsic muscles attach within the dorsal compartment, there are only three intrinsic muscles which act on the foot and two extrinsic muscles which act on the foot rather than the ankle.

- Extensor Digitorum Longus: The extensor digitorum longus is a deep-lying extrinsic muscle that lies runs down the length of the tibia.
  - Attachments: Originates from the tibia and transitions into a tendon, passes into the foot, splits into four, and attaches to the toes.
  - Actions: Extension of the toes.
- Extensor Digitorum Brevis: The extensor digitorum brevis muscle is a deep-lying intrinsic muscle lying beneath the tendon of the extensor digitorum longus.
  - Attachments: Originates from the heel and attaches to the toes.
  - Actions: Extension of the toes.

- Extensor Hallucis Longus: The extensor hallucis longus is a deep-lying extrinsic muscle lying beneath the extensor digitorum longus.
    - Attachments: Originates from the fibula and attaches to the big toe.
    - Actions: Extension of the big toe.
- Extensor Hallucis Brevis: The extensor hallucis brevis muscle is an intrinsic muscle of the foot lying between the extensor digitorum longus and lateral to extensor hallucis longus.
    - Attachments: Originates from the heel and attaches to the big toe.
    - Actions: Extension of the big toe.
- Dorsal Interossei: There are four dorsal interossei located between the metatarsals. Each arises from two metatarsals.
    - Attachments: Originates from the sides of metatarsals one to five (big toe to little toe). The first muscle attaches to the medial side of the phalanx of the second toe. The second to fourth interossei attach to the lateral sides of the phalanxes of toes two to four.
    - Actions: Abduct and flexes digits two to four.

Plantar Compartment

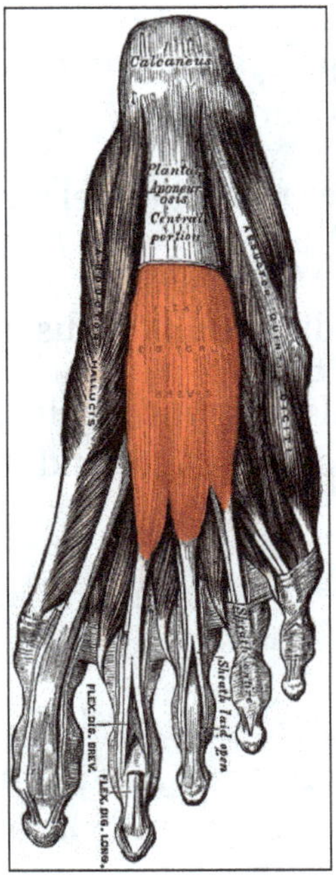

Plantar View of Foot: The muscles, tendons, and ligaments of the sole of the foot with flexor digitorum brevis shown in red.

The muscles of the plantar compartment play a key role in stabilizing the arch of the foot and controlling the digits, providing the necessary strength and fine movements to maintain balance and posture and promote walking.

- Abductor Hallucis: The abductor hallucis muscle is located on the medial side of the sole.
  - Attachments: Originates from the heel and attaches to the base of the big toe.
  - Actions: Abducts and flexes the big toe.
- Flexor Digitorum Brevis: The flexor digitorum brevis muscle is located laterally to the abductor hallucis and lies in the middle of the sole of the foot.
  - Attachments: Originates from the heel and attaches to all the toes excluding the little toe.
  - Actions: Flexes the toes at the proximal interphalangeal (between the phalanges preoxmales and phalanges mediae) joints.
- Abductor Digiti Minimi: The abductor digiti minimi muscle is located on the lateral side of the foot.
  - Attachments: Originates from the heel and attaches to the little toe.
  - Actions: Abducts and flexes the little toe.
- Quadratus Plantae: The quadratus plantae muscle is located superior to the flexor digitorum longus tendons.
  - Attachments: Originates from the heel and attaches to the tendons of flexor digitorum longus.
  - Actions: Assists flexor digitorum longus in flexing the lateral four toes.
- Lumbricals: The lumbricals are four small skeletal muscles, accessory to the tendons of the flexor digitorum longus and numbered from the medial side of the foot.
  - Attachments: Originates from the tendons of flexor digitorum longus. Attaches to the toes.
  - Actions: Flexes at the metatarsophalangeal (between the metatarsals and phalanges proximale) joints, while extending the interphalangeal joints (between the phalanges proximale and phalanges mediae).
- Flexor Hallucis Brevis: The flexor hallucis brevis muscle is located on the medial side of the foot.

- Attachments: Originates from the plantar surfaces of the cuboid and lateral cuneiform bones in the sole of the foot and attaches to the big toe.
- Actions: Flexes the big toe.

- Adductor Hallucis: The adductor hallucis muscle is located laterally to the flexor hallucis brevis.
  - Attachments: The adductor hallucis originates from the bases of the first four metatarsals, and also from the plantar ligaments before attaching to the big toe.
  - Actions: Adduct the big toe and maintains the arch of the foot.
- Plantar Interossei: The three plantar interossei lie beneath rather than between the metatarsal bones. Each is connected with one metatarsal bone.
  - Attachments: Originates from the medial side of metatarsals three to five. Attaches to the medial sides of the phalanges of digits three to five.
  - Actions: Adducts and flexes digits three to five.
- Flexor Digiti Minimi Brevis: Similar to the interossei in structure, the flexor digiti minimi brevis muscle is located on the lateral side of the foot underneath the metatarsal of the little toe.
  - Attachments: Originates from the base of the fifth metatarsal and attaches to the base of phalanx of the little toe.
  - Actions: Flexes the little toe.

References

- Foot-anatomy-and-physiology-3119204: verywellhealth.com, Retrieved 09 June, 2020
- Burgener, Francis A.; Meyers, Steven P.; Tan, Raymond K.; Zaunbauer, Wolfgang (2002). Differential Diagnosis in Magnetic Resonance Imaging. Thieme. ISBN 978-1-58890-085-2., p. 390
- Anatomy-ankle, ankle-doctor: scoi.com, Retrieved 10 July, 2020
- Gottsegen, CJ; Eyer, BA; White, EA; Learch, TJ; Forrester, D (2008). "Avulsion fractures of the knee: imaging findings and clinical significance". Radiographics. 28 (6): 1755–1770. doi:10.1148/rg.286085503. PMID 18936034
- Obturator-foramen, anatomy-feed: anatomyzone.com, Retrieved 30 June, 2020

Permissions

All chapters in this book are published with permission under the Creative Commons Attribution Share Alike License or equivalent. Every chapter published in this book has been scrutinized by our experts. Their significance has been extensively debated. The topics covered herein carry significant information for a comprehensive understanding. They may even be implemented as practical applications or may be referred to as a beginning point for further studies.

We would like to thank the editorial team for lending their expertise to make the book truly unique. They have played a crucial role in the development of this book. Without their invaluable contributions this book wouldn't have been possible. They have made vital efforts to compile up to date information on the varied aspects of this subject to make this book a valuable addition to the collection of many professionals and students.

This book was conceptualized with the vision of imparting up-to-date and integrated information in this field. To ensure the same, a matchless editorial board was set up. Every individual on the board went through rigorous rounds of assessment to prove their worth. After which they invested a large part of their time researching and compiling the most relevant data for our readers.

The editorial board has been involved in producing this book since its inception. They have spent rigorous hours researching and exploring the diverse topics which have resulted in the successful publishing of this book. They have passed on their knowledge of decades through this book. To expedite this challenging task, the publisher supported the team at every step. A small team of assistant editors was also appointed to further simplify the editing procedure and attain best results for the readers.

Apart from the editorial board, the designing team has also invested a significant amount of their time in understanding the subject and creating the most relevant covers. They scrutinized every image to scout for the most suitable representation of the subject and create an appropriate cover for the book.

The publishing team has been an ardent support to the editorial, designing and production team. Their endless efforts to recruit the best for this project, has resulted in the accomplishment of this book. They are a veteran in the field of academics and their pool of knowledge is as vast as their experience in printing. Their expertise and guidance has proved useful at every step. Their uncompromising quality standards have made this book an exceptional effort. Their encouragement from time to time has been an inspiration for everyone.

The publisher and the editorial board hope that this book will prove to be a valuable piece of knowledge for students, practitioners and scholars across the globe.

Index

www.ingramcontent.com/pod-product-compliance
Lightning Source LLC
LaVergne TN
LVHW080628210926
848485LV00003B/227
* 9 7 8 1 6 3 9 8 7 3 1 1 1 *